Review of the U.S. Navy Environmental Health Center's Health-Hazard Assessment Process

Subcommittee on Toxicological Hazard and Risk Assessment

Committee on Toxicology

Board on Environmental Studies and Toxicology

Commission on Life Sciences

National Research Council

NATIONAL ACADEMY PRESS
Washington, D.C.

NATIONAL ACADEMY PRESS   2101 Constitution Ave., N.W.   Washington, D.C. 20418

NOTICE: The project that is the subject of this report was approved by the Governing Board of the National Research Council, whose members are drawn from the councils of the National Academy of Sciences, the National Academy of Engineering, and the Institute of Medicine. The members of the committee responsible for the report were chosen for their special competences and with regard for appropriate balance.

This project was supported by Contract Nos. DAMD17-89-C-9086 and DAMD17-99-C-9049 between the National Academy of Sciences and U.S. Army. Any opinions, findings, conclusions, or recommendations expressed in this publication are those of the author(s) and do not necessarily reflect the view of the organizations or agencies that provided support for this project.

International Standard Book Number 0-309-07147-X

Additional copies of this report are available from:

National Academy Press
2101 Constitution Ave., NW
Box 285
Washington, DC 20055

800-624-6242
202-334-3313 (in the Washington metropolitan area)
http://www.nap.edu

Copyright 2000 by the National Academy of Sciences. All rights reserved.

Printed in the United States of America

THE NATIONAL ACADEMIES

National Academy of Sciences
National Academy of Engineering
Institute of Medicine
National Research Council

The **National Academy of Sciences** is a private, nonprofit, self-perpetuating society of distinguished scholars engaged in scientific and engineering research, dedicated to the furtherance of science and technology and to their use for the general welfare. Upon the authority of the charter granted to it by the Congress in 1863, the Academy has a mandate that requires it to advise the federal government on scientific and technical matters. Dr. Bruce M. Alberts is president of the National Academy of Sciences.

The **National Academy of Engineering** was established in 1964, under the charter of the National Academy of Sciences, as a parallel organization of outstanding engineers. It is autonomous in its administration and in the selection of its members, sharing with the National Academy of Sciences the responsibility for advising the federal government. The National Academy of Engineering also sponsors engineering programs aimed at meeting national needs, encourages education and research, and recognizes the superior achievements of engineers. Dr. William A. Wulf is president of the National Academy of Engineering.

The **Institute of Medicine** was established in 1970 by the National Academy of Sciences to secure the services of eminent members of appropriate professions in the examination of policy matters pertaining to the health of the public. The Institute acts under the responsibility given to the National Academy of Sciences by its congressional charter to be an adviser to the federal government and, upon its own initiative, to identify issues of medical care, research, and education. Dr. Kenneth I. Shine is president of the Institute of Medicine.

The **National Research Council** was organized by the National Academy of Sciences in 1916 to associate the broad community of science and technology with the Academy's purposes of furthering knowledge and advising the federal government. Functioning in accordance with general policies determined by the Academy, the Council has become the principal operating agency of both the National Academy of Sciences and the National Academy of Engineering in providing services to the government, the public, and the scientific and engineering communities. The Council is administered jointly by both Academies and the Institute of Medicine. Dr. Bruce M. Alberts and Dr. William A. Wulf are chairman and vice chairman, respectively, of the National Research Council.

Subcommittee on Toxicological Hazard and Risk Assessment

ROGENE HENDERSON *(CHAIR)*, Lovelace Respiratory Research Institute, Albuquerque, New Mexico
GERMAINE M. BUCK, University at Buffalo, State of New York
JACK H. DEAN, Sanofi-Synthelabo, Inc., Malvern, Pennsylvania
KEVIN E. DRISCOLL, Proctor and Gamble Pharmaceuticals, Cincinnati, Ohio
DAVID W. GAYLOR, U.S. Food and Drug Administration, Jefferson, Arkansas
JUDITH A. GRAHAM, U.S. Environmental Protection Agency, Research Triangle Park, North Carolina
JOHN L. O'DONOGHUE, Eastman Kodak Company, Rochester, New York
ROBERT SNYDER, Rutgers University, Piscataway, New Jersey
BERNARD M. WAGNER, Consultant, Short Hills, New Jersey
ANNETTA P. WATSON, Oak Ridge National Laboratory, Oak Ridge, Tennessee
HANSPETER R. WITSCHI, University of California, Davis, California

Staff

KULBIR S. BAKSHI, Project Director
ROBERT J. CROSSGROVE, Editor
MIRSADA KARALIC-LONCAREVIC, Information Specialist
TANYA LEE, Acting Project Assistant

Sponsor

U.S. DEPARTMENT OF DEFENSE

COMMITTEE ON TOXICOLOGY

BAILUS WALKER, JR. *(Chair)*, Howard University Medical Center and American Public Health Association, Washington, D.C.
MELVIN E. ANDERSEN, Colorado State University, Denver, Colorado
GERMAINE M. BUCK, University at Buffalo, State of New York
ROBERT E. FORSTER II, University of Pennsylvania, Philadelphia, Pennsylvania
PAUL M.D. FOSTER, Chemical Industry Institute of Toxicology, Research Triangle Park, North Carolina
WILLIAM E. HALPERIN, National Institute for Occupational Safety and Health, Cincinnati, Ohio
CHARLES H. HOBBS, Lovelace Respiratory Research Institute and Lovelace Biomedical and Environmental Research Institute, Albuquerque, New Mexico
SAM KACEW, Department of Pharmacology Faculty of Medicine and University of Ottawa, Ontario, Canada
NANCY KERKVLIET, Oregon State University, Agricultural and Life Sciences, Corvallis, Oregon
FLORENCE K. KINOSHITA, Hercules Incorporated, Wilmington, Delaware
MICHAEL J. KOSNETT, University of Colorado Health Sciences Center, Denver, Colorado
MORTON LIPPMANN, New York University School of Medicine, Tuxedo, New York
ERNEST E. MCCONNELL, ToxPath, Inc., Raleigh, North Carolina
THOMAS E. MCKONE, Lawrence Berkeley National Laboratory and University of California, Berkeley, California
HARIHARA MEHENDALE, The University of Louisiana at Monroe, Monroe, Louisiana
DAVID H. MOORE, Battelle Memorial Institute, Bel Air, Maryland
GÜNTER OBERDÖRSTER, University of Rochester, Rochester, New York
JOHN L. O'DONOGHUE, Eastman Kodak Company, Rochester, New York
GEORGE M. RUSCH, AlliedSignal, Inc., Morristown, New Jersey
MARY E. VORE, University of Kentucky, Lexington, Kentucky
ANNETTA P. WATSON, Oak Ridge National Laboratory, Oak Ridge, Tennessee
LAUREN ZEISE, Office of Environmental Health Hazard Assessment, Oakland, California

Staff

KULBIR S. BAKSHI, Program Director
SUSAN N.J. PANG, Program Officer
ABIGAIL E. STACK, Program Officer
KATHRINE J. IVERSON, Manager, Toxicology Information Center
TANYA LEE, Acting Project Assistant
EMILY SMAIL, Project Assistant

BOARD ON ENVIRONMENTAL STUDIES AND TOXICOLOGY

GORDON ORIANS (*Chair*), University of Washington, Seattle, Washington
DONALD MATTISON (*Vice Chair*), March of Dimes, White Plains, New York
DAVID ALLEN, University of Texas, Austin, Texas
INGRID C. BURKE, Colorado State University, Fort Collins, Colorado
WILLIAM L. CHAMEIDES, Georgia Institute of Technology, Atlanta, Georgia
JOHN DOULL, University of Kansas Medical Center, Kansas City, Kansas
CHRISTOPHER B. FIELD, Carnegie Institute of Washington, Stanford, California
JOHN GERHART, University of California, Berkeley, California
J. PAUL GILMAN, Celera Genomics, Rockville, Maryland
BRUCE D. HAMMOCK, University of California, Davis, California
MARK HARWELL, University of Miami, Miami, Florida
ROGENE HENDERSON, Lovelace Respiratory Research Institute, Albuquerque, New Mexico
CAROL HENRY, Chemical Manufacturers Association, Arlington, Virginia
BARBARA HULKA, University of North Carolina, Chapel Hill, North Carolina
JAMES F. KITCHELL, University of Wisconsin, Madison, Wisconsin
DANIEL KREWSKI, University of Ottawa, Ottawa, Ontario
JAMES A. MACMAHON, Utah State University, Logan, Utah
MARIO J. MOLINA, Massachusetts Institute of Technology, Cambridge, Massachusetts
CHARLES O'MELIA, Johns Hopkins University, Baltimore, Maryland
WILLEM F. PASSCHIER, Health Council of the Netherlands
KIRK SMITH, University of California, Berkeley, California
MARGARET STRAND, Oppenheimer Wolff Donnelly & Bayh, LLP, Washington, D.C.
TERRY F. YOSIE, Chemical Manufacturers Association, Arlington, Virginia

Senior Staff

JAMES J. REISA, Director
DAVID J. POLICANSKY, Associate Director and Senior Program Director for Applied Ecology
CAROL A. MACZKA, Senior Program Director for Toxicology and Risk Assessment
RAYMOND A. WASSEL, Senior Program Director for Environmental Sciences and Engineering
KULBIR BAKSHI, Program Director for the Committee on Toxicology
LEE R. PAULSON, Program Director for Resource Management
ROBERTA M. WEDGE, Program Director for Risk Analysis

COMMISSION ON LIFE SCIENCES

MICHAEL T. CLEGG *(Chair)*, University of California, Riverside, California
PAUL BERG *(Vice Chair)*, Stanford University, Stanford, California
FREDERICK R. ANDERSON, Cadwalader, Wickersham & Taft, Washington, D.C.
JOANNA BURGER, Rutgers University, Piscataway, New Jersey
JAMES E. CLEAVER, University of California, San Francisco, California
DAVID S. EISENBERG, University of California, Los Angeles, California
JOHN L. EMMERSON, Fishers, Indiana
NEAL L. FIRST, University of Wisconsin, Madison, Wisconsin
DAVID J. GALAS, Keck Graduate Institute of Applied Life Science, Claremont, California
DAVID V. GOEDDEL, Tularik, Inc., South San Francisco, California
ARTURO GOMEZ-POMPA, University of California, Riverside, California
COREY S. GOODMAN, University of California, Berkeley, California
JON W. GORDON, Mount Sinai School of Medicine, New York, New York
DAVID G. HOEL, Medical University of South Carolina, Charleston, South Carolina
BARBARA S. HULKA, University of North Carolina, Chapel Hill, North Carolina
CYNTHIA J. KENYON, University of California, San Francisco, California
BRUCE R. LEVIN, Emory University, Atlanta, Georgia
DAVID M. LIVINGSTON, Dana-Farber Cancer Institute, Boston, Massachusetts
DONALD R. MATTISON, March of Dimes, White Plains, New York
ELLIOT M. MEYEROWITZ, California Institute of Technology, Pasadena, California
ROBERT T. PAINE, University of Washington, Seattle, Washington
RONALD R. SEDEROFF, North Carolina State University, Raleigh, North Carolina
ROBERT R. SOKAL, State University of New York, Stony Brook, New York
CHARLES F. STEVENS, The Salk Institute for Biological Studies, La Jolla, California
SHIRLEY M. TILGHMAN, Princeton University, Princeton, New Jersey
RAYMOND L. WHITE, University of Utah, Salt Lake City, Utah

Staff

WARREN R. MUIR, Executive Director
JACQUELINE K. PRINCE, Financial Officer
BARBARA B. SMITH, Administrative Associate
LAURA T. HOLLIDAY, Senior Program Assistant

OTHER REPORTS OF THE
BOARD ON ENVIRONMENTAL STUDIES AND TOXICOLOGY

Strengthening Science at the U.S. Environmental Protection Agency: Research Management and Peer Review Practice (2000)
Scientific Frontiers in Developmental Toxicology and Risk Assessment (2000)
Modeling Mobile-Source Emissions (2000)
Copper in Drinking Water (2000)
Ecological Indicators for the Nation (2000)
Waste Incineration and Public Health (1999)
Hormonally Active Agents in the Environment (1999)
Research Priorities for Airborne Particulate Matter: I. Immediate Priorities and a Long-Range Research Portfolio (1998); II. Evaluating Research Progress and Updating the Portfolio (1999)
Ozone-Forming Potential of Reformulated Gasoline (1999)
Risk-Based Waste Classification in California (1999)
Arsenic in Drinking Water (1999)
Brucellosis in the Greater Yellowstone Area (1998)
The National Research Council's Committee on Toxicology: The First 50 Years (1997)
Toxicologic Assessment of the Army's Zinc Cadmium Sulfide Dispersion Tests (1997)
Carcinogens and Anticarcinogens in the Human Diet (1996)
Upstream: Salmon and Society in the Pacific Northwest (1996)
Science and the Endangered Species Act (1995)
Wetlands: Characteristics and Boundaries (1995)
Biologic Markers (5 reports, 1989-1995)
Review of EPA's Environmental Monitoring and Assessment Program (3 reports, 1994-1995)
Science and Judgment in Risk Assessment (1994)
Ranking Hazardous Waste Sites for Remedial Action (1994)
Pesticides in the Diets of Infants and Children (1993)
Issues in Risk Assessment (1993)
Setting Priorities for Land Conservation (1993)
Protecting Visibility in National Parks and Wilderness Areas (1993)
Dolphins and the Tuna Industry (1992)
Hazardous Materials on the Public Lands (1992)
Science and the National Parks (1992)
Animals as Sentinels of Environmental Health Hazards (1991)
Assessment of the U.S. Outer Continental Shelf Environmental Studies Program, Volumes I-IV (1991-1993)
Human Exposure Assessment for Airborne Pollutants (1991)
Monitoring Human Tissues for Toxic Substances (1991)
Rethinking the Ozone Problem in Urban and Regional Air Pollution (1991)
Decline of the Sea Turtles (1990)

*Copies of these reports may be ordered from
the National Academy Press
(800) 624-6242 or (202) 334-3313
www.nap.edu*

OTHER REPORTS OF THE COMMITTEE ON TOXICOLOGY

Review of the U.S. Navy's Exposure Standard for Manufactured Vitreous Fibers (2000)
Submarine Exposure Guidance Levels for Selected Hydrofluorocarbons: HFC-236fa, HFC-23, and HFC-404a (2000)
Health Risk Assessment of Selected Flame-Retardant Chemicals (2000)
Review of the U.S. Army's Health Risk Assessments for Oral Exposure to Six Chemical-Warfare Agents (1999)
Toxicity of Military Smokes and Obscurants, Volume 1 (1997), Volume 2 (1999), Volume 3 (1999)
Assessment of Exposure-Response Functions for Rocket-Emission Toxicants (1998)
Review of a Screening Level Risk Assessment for the Naval Air Facility at Atsugi, Japan (Letter Report) (1998)
Toxicity of Alternatives to Chlorofluorocarbons: HFC-134a and HCFC-123 (1996)
Permissible Exposure Levels for Selected Military Fuel Vapors (1996)
Spacecraft Maximum Allowable Concentrations for Selected Airborne Contaminants, Vol. 1 (1994), Vol. 2 (1996), Vol. 3 (1996), Vol. 4 (2000)

Preface

THE U.S. NAVY Environmental Health Center (NEHC) supports the preventive medicine program of the Navy, especially in the areas of occupational health and public health. NEHC receives numerous requests to evaluate potential health hazards associated with materials used by the Navy and Marine Corps. In response to such requests, NEHC develops and reviews toxicological and related data and makes recommendations of acceptable exposures to these materials based on their potential to produce toxic effects in humans.

As part of its efforts to protect Navy personnel and their families from exposures to toxic chemicals, the Navy's Office of the Surgeon General asked the National Research Council (NRC) to independently review the adequacy of the NEHC health- hazard assessment (HHA) process. The NRC assigned this task to the Committee on Toxicology (COT) of the Board on Environmental Studies and Toxicology. The COT established the Subcommittee on Toxicological Hazard Evaluation, which prepared this report.

The subcommittee was asked to assess the validity and effectiveness of NEHC's HHA process; to determine whether the process as implemented provides the Navy with state-of-the-art, comprehensive, and defensible evaluations of toxicological hazards; and to identify any program elements that require improvement. This report is intended to provide NEHC with recommendations that will improve and strengthen the HHA process and aid the Navy's efforts related to preventive medicine.

The subcommittee gratefully acknowledges Capt. David Macys, Capt.

Richard Buck, Commander William Luttrell, Capt. George Kramer, Capt. Kenneth Still, James Crawl, Gerald Drewyer, Andrea Lunsford, Vera Wang, Charles Gross, Steven Sorgen (all from the U.S. Navy), and Dr. Ronald Wolff (Lilly Research Laboratories) for providing background information and for making presentations to the subcommittee.

This report has been reviewed by individuals chosen for their diverse perspectives and technical expertise in accordance with procedures for reviewing NRC reports approved by the NRC's Report Review Committee. The purpose of this independent review was to provide candid and critical comments to assist the NRC in making the published report as sound as possible and to ensure that the report meets institutional standards for objectivity, evidence, and responsiveness to the study charge. The review comments and draft manuscript remain confidential to protect the integrity of the deliberative process. We wish to thank the following individuals, who are neither officials nor employees of the NRC, for their participation in the review of this report: Sidney Green, Howard University; George Rusch, AlliedSignal, Inc.; Donald Gardner, Inhalation Toxicology Associates; Joseph Barzelleca, Virginia Commonwealth University, and John Doull, University of Kansas Medical Center.

The individuals listed above have provided many constructive comments and suggestions. It may be emphasized, however, that responsibility for the final content of this report rests entirely with the authoring committee and the NRC.

The subcommittee was ably guided and assisted by staff of the NRC's Board on Environmental Studies and Toxicology, especially Kulbir S. Bakshi (project director), Robert Crossgrove (editor), Evelyn Simeon, and Pamela Friedman (project administrative assistants). These staff members merit special recognition for their thoughtful contributions and extraordinary efforts in producing the report.

Finally, we would like to express my thanks and admiration to the members of the subcommittee for their dedicated efforts throughout the development of the report.

Rogene Henderson, *Chair*
Subcommittee on Toxicological Hazard
and Risk Assessment

Bailus Walker, *Chair*
Committee on Toxicology

Contents

ABBREVIATIONS xv

EXECUTIVE SUMMARY 1

1 INTRODUCTION 11
   Background, 12
   Subcommittee's Approach to the Charge, 13
   Structure of this Report, 14

2 THE NAVY'S CURRENT HEALTH-HAZARD
   ASSESSMENT PROCESS 15
   Navy Policies and Directives Related to Hazardous
      Substance,15
   Navy's Current Health-Hazard Assessment Process, 19
   NEHC's Process for Preparing HHA Reports, 24
   Special Considerations for Risk Assessments:
      Populations at Risk, 31
   Exposure Information for Health-Hazard Assessments, 32
   Human Resources Available to Conduct Health-Hazard
      Assessments, 33
   Information Systems Used for Hazard Assessments, 34
   Peer Review of Reports, 35

xiii

3 OTHER RELATED HEALTH-HAZARD ASSESSMENT PROCESSES ... 37
Consumer Product and Pharmaceutical Industries, 37
U.S. Army Center for Health Promotion and Preventive Medicine, 41
Recommendation, 44

4 CONCLUSIONS AND RECOMMENDATIONS ... 45
Documentation and Development of Standard Operating Procedures, 46
Staffing, 47
Data Acquisition and Management, 50
Information Sources for Conducting Health-Hazard Assessments, 51
Quality Assurance and Quality Control, 53
Communication Within Other Navy Programs and With Other Organizations, 54
Medical Surveillance and Centralization of Medical Data, 56
Life-Cycle Assessment, 56
Overall Summary, 57

REFERENCES ... 58

APPENDIX A
History of NEHC and Its Relationship with Other Navy Organizations ... 63

APPENDIX B
Department of Defense and Navy Directives and Regulations Relating to the Use of Hazardous Materials . 67

Abbreviations

AAALAC	American Association for the Accreditation of Laboratory Animal Care
ACGIH	American Conference of Governmental Industrial Hygienists
AEGL	acute exposure guideline level
AEHA	U.S. Army Environmental Hygiene Agency (now known as CHPPM)
AIHA	American Industrial Hygiene Association
ATSDR	Agency for Toxic Substances and Disease Registry
BUMED	Bureau of Medicine and Surgery
CEGL	continuous exposure guidance level
CHPPM	U.S. Army Center for Health Promotion and Preventive Medicine
CNO	Chief of Naval Operations
COT	Committee on Toxicology
DOD	Department of Defense
DON	Department of the Navy
DTIC	Defense Technical Information Center
EEGL	emergency exposure guidance level
EPA	U.S. Environmental Protection Agency
GLP	good laboratory practice
HCS	hazard communication standard

HEAST	health effects assessment summary table
HHA	health hazard assessment
HMD	Hazardous Materials Department
HSDB	Hazardous Substances Data Base
IARC	International Agency for Research on Cancer
IERA	Institute for Environment, Safety, and Occupational Health Risk Analysis
IH	industrial hygiene
IPCS	International Program on Chemical Safety
IRIS	Integrated Risk Information System
ISO	International Organization for Standardization
LCA	life-cycle assessment
MSDS	Material Safety Data Sheet
NAVOSH	Navy Occupational Safety and Health
NEHC	Navy Environmental Health Center
NHRC	Naval Health Research Center
NIOSH	National Institute for Occupational Safety and Health
NHRC/TD	Naval Health Research Center's Toxicology Detachment
NRC	National Research Council
ORNL	Oak Ridge National Laboratory
OSHA	Occupational Safety and Health Administration
QA/QC	quality assurance/quality control
QAO	Quality Assurance Office
RTECS	Registry of Toxic Effects
SOP	standard operating procedure
SECNAV	Secretary of the Navy
SMAC	spacecraft maximum allowable concentration
WPAFB	Wright-Patterson Air Force Base

Review of the U.S. Navy Environmental Health Center's Health-Hazard Assessment Process

Executive Summary

THE NAVY AND MARINE CORPS use a large number of chemicals on land at shore facilities, in the air in combat and reconnaissance aircraft, on seas around the world in surface vessels, and in submarine vessels that operate as self-contained environments. Although many of the chemicals used by the Navy might be relatively innocuous, the Navy does use a large number that can pose significant health hazards under specific exposure circumstances.

The Navy Environmental Health Center (NEHC) is the primary organization within the Navy that is tasked with assessing occupational and environmental health hazards for Navy personnel from exposures to toxic substances. It serves as the central source that provides the Navy and Marine Corps, ashore and afloat, with technical support for preventive medicine, medical management, health promotion, drug screening, and occupational and environmental health programs. For many of these programs, NEHC reviews toxicological and related data and prepares health-hazard assessments (HHAs) for potentially hazardous materials under a variety of exposure conditions. Because NEHC is continually being asked to develop HHAs for the Navy and the Marine Corps, and because the Navy is committed to protecting its personnel from exposures to toxic chemicals, the National Research Council (NRC) was asked to assess independently the validity and effectiveness of NEHC's HHA process[1]; to determine whether the process as implemented provides the Navy with the state-of-the-art, comprehensive, and defensible evaluations of health hazards; and to identify any program elements that require improvement.

The NRC assigned this project to the Board on Environmental Studies and Toxicology's Committee on Toxicology (COT). COT convened the Subcommittee on Toxicological Hazard and Risk Assessment, which prepared this report. The subcommittee has expertise in general toxicology, inhalation toxicology, epidemiology, neurotoxicology, immunotoxicology, reproductive and developmental toxicology, pharmacology, medicine, risk assessment, and biostatistics.

THE SUBCOMMITTEE'S APPROACH TO ITS CHARGE

The subcommittee's assessment of NEHC's HHA process is based on its review of documents submitted by NEHC; presentations made by NEHC personnel at subcommittee meetings; and site visits to NEHC in Norfolk, Virginia, and the aircraft carrier, U.S.S. *Constellation*, while docked at the Naval Air Station North Island, San Diego, California. In addition, the subcommittee reviewed the HHA processes used by some chemical and pharmaceutical companies for their adaptability and usefulness to the Navy's situation.

CONCLUSIONS AND RECOMMENDATIONS

The subcommittee has reviewed NEHC's HHA process and concludes that NEHC has generally done an adequate job preparing routine HHAs, considering the NEHC's available resources. Several deficiencies are noted, however, especially for conducting complex HHAs. The deficiencies include (1) the lack of formal, written, standard operating procedures (SOPs) for preparing HHAs, (2) inadequate in-house staff expertise for preparing complex HHAs, (3) inadequate availability

---

[1] HHAs are conducted by the Industrial Hygiene Directorate's Hazardous Materials Department (HMD) of NEHC.

of electronic databases, (4) inadequate quality-assurance and quality-control procedures, (5) inadequate coordination and information transfer between NEHC and other stakeholders, and (6) inadequate medical surveillance as well as the absence of a centralized medical-data management structure. The subcommittee's conclusions and recommendations with respect to each of these deficiences are discussed below.

Documentation and Development of Standard Operating Procedures

In reviewing the NEHC's procedures for conducting HHAs, it became apparent to the subcommittee that no formal procedures (e.g., SOPs, including flow charts) have been developed as to how HHAs should be prepared and documented. To improve procedures currently used by NEHC, the subcommittee recommends that NEHC utilize procedures established in industry (e.g., pharmaceutical and chemical companies), governmental agencies, and other organizations. The subcommittee recommends that NEHC develop a set of SOPs for the preparation of its HHAs by incorporating the relevant aspects of procedures employed by those groups.

The subcommittee also recommends that the NEHC develop guidelines or criteria for developing HHAs or for deferring a review to NEHC, for use by industrial hygiene personnel on ships or at regional occupational health departments.

Staffing

The effectiveness of the NEHC's HHA program is dependent on the training and expertise of the personnel tasked to develop HHAs. The subcommittee believes that much of the work performed by the NEHC can be carried out by scientists or industrial hygienists at the bachelor or master of science level. The subcommittee concludes that the current education and experience level of NEHC staff is adequate for preparing routine HHAs. However, there were a few complex risk-assessment projects, such as those that involved determining the health hazards associated with off-gassing of chemicals in submarines, which required personnel that are more highly trained in toxicology, industrial hygiene, epidemiology, and human risk assessment. The subcommittee recommends that NEHC recruit additional scientists with expertise in toxicology, epidemiology, and risk assessment for conducting such complex tasks. The subcommittee also recommends that all naval operations handling hazardous chemical substances should have an adequate level of access to industrial hygiene personnel. This expertise needs to be commensurate with the size of the facility. For example, large facilities (such as an aircraft carrier that accommodates up to 5,000 naval personnel) should have more than one industrial hygienist to support the continuous or sustained operations typical for deploying naval vessels, and in case one of the officers becomes ill, injured, or transferred. Small facilities, such as a submarine, on the other hand, would only need periodic access to such personnel.

Because of budget reductions in the Navy, the combination of decreasing numbers of experienced staff and an increasing demand for greater number of HHAs requires development of a more effective approach for conducting HHAs. The subcommittee recommends that the NEHC develop a long-term strategy to deal with increasing demand for services in the face of decreasing resources. This strategy would include elements such as streamlined processes to conserve staff time; increased training of current staff to keep up to date with advances in toxicology and risk assessment; and development of a workforce planning strategy that would include a succession plan for NEHC staff and a projection of future personnel needs, along with minimal training and experience requirements for each position. Particular attention should be given to the qualifications necessary for personnel exercising key technical oversight review function for HHAs.

Data Acquisition and Management

Based on its review of the information sources currently available to key NEHC staff for conducting HHAs, the subcommittee concluded that there is an absence or limited availability of computerized hardware and software for accessing electronic information databases. Routine access to such databases is needed to ensure that the most up-to-date information is obtained for preparing HHAs. The absence of a data-management structure also impedes meaningful analysis of the vast array of existing data available throughout the Navy on chemicals, exposures, and health outcomes. The subcommittee recommends that NEHC staff be provided with, and trained to use, up-to-date computer hardware and software for conducting electronic searches. In addition, the subcommittee recommends that NEHC develop a literature-search strategy for obtaining the most up-to-date information. In-depth literature searches should be performed for new or experimental compounds or substances, whereas less comprehensive searches are needed for chemicals or substances that are in common use (e.g., cleaning supplies and certain paints).

NEHC currently relies heavily on SmartRisk/SmartTox® assessment software and Material Safety Data Sheets (MSDSs) when conducting HHAs. The subcommittee recommends that NEHC not rely solely on these sources. Although it is appropriate to begin an assessment with consideration of the MSDSs, all data contained in them should be independently confirmed before their use. SmartRisk/SmartTox® software is of limited usefulness when evaluating the toxicity of compounds, because the information contained in it might not be current and must be updated periodically.

There are a number of highly credible information sources on the toxicology of industrial and commercial compounds that could provide valuable technical input in this context. Many are already used to some extent by the NEHC HHA staff. Examples include the U.S. Environmental Protection Agency (EPA) Integrated Risk Information System, Health Effects Assessment Summary Tables, the Hazardous Substances Data Base (HSDB), Registry of Toxic Effects, the *Agrochemicals Handbook (Royal Society of Chemistry 1994)*, *Sax's Dangerous Properties of Industrial Materials (Lewis 1996)*, *Patty's Industrial Hygiene and Toxicology* (Clayton and Clayton 1993), the International Agency for Research on Cancer (IARC) series, and the Air Force toxicology guide *Installation Restoration Program Toxicology Guide* (ORNL 1989, 1990). The EPA *Exposure Factors Handbook* can be consulted for updated exposure factors.

To avoid duplication, NEHC should explore the use of additional authoritative sources, such as existing hazard and risk assessments conducted by other Department of Defense (DOD) and governmental organizations, private organizations, and those available in the open literature. Examples include acute exposure guideline levels (AEGLs) developed by the National Advisory Committee on AEGLs (these documents are also reviewed by COT), carcinogenicity evaluations prepared by IARC, and documents prepared by COT such as spacecraft maximum allowable concentrations (SMACs), emergency exposure guidance levels (EEGLs), and continuous exposure guidance levels (CEGLs). For repeated exposures, the Navy should also routinely review threshold limit values proposed by the American Conference of Governmental Industrial Hygienists, and the workplace environmental exposure limits proposed by the American Industrial Hygiene Association.

Quality Assurance and Quality Control

The subcommittee concludes that the NEHC's HHA program has inadequate formal quality-assurance and quality-control (QA/QC) procedures. The subcommittee recommends that NEHC establish a QA/QC program to (1) review and maintain updated SOPs for developing HHAs, and (2) ensure that HHA documents developed by NEHC, staff, and contractors are scientifically sound and instructive.

To ensure the scientific accuracy of HHA reports, the subcommittee recommends that a system be developed for regular peer review of HHAs by qualified internal and external reviewers. This system should include criteria for determining whether an HHA would undergo external or internal review and what types of expertise and institutions are needed to perform such reviews, and for documenting the process and its results.

This review would help to ensure scientific rigor and objectivity and provide an opportunity for staff to obtain additional perspective from scientists outside the Navy. Furthermore, the subcommittee recommends that a peer review board be established to provide a periodic external review of NEHC's HHA process. The board should be an independent body comprised of scientists possessing experience in industrial hygiene, toxicology, and risk assessment.

Communication Within the Navy and With Other Organizations

The subcommittee observed that there is little or no communication between the NEHC's HHA program and its clients in various Navy commands. The subcommittee recommends that the NEHC establish and implement systematic customer survey and feedback mechanisms to determine utility and timeliness of its HHAs in decision making, and to obtain suggestions for improvement to better serve the needs of its clients or customers (e.g., Naval Sea Systems Command).

The subcommittee believes that there is a need for greater coordination and information transfer between the NEHC and other Navy or governmental bodies that also perform HHAs. This interaction could provide insight for addressing issues and solving problems that may be common between institutions. NEHC would benefit by interacting more with the U.S. Army Center for Health Promotion and Preventive Medicine; the U.S. Air Force's Institute for Environmental Safety and Occupational Health Risk Analysis at Brooks Air Force Base, Texas; Triservice Toxicology Research Laboratories at Wright-Patterson Air Force Base in Ohio; United States Environmental Protection Agency; Occupational Safety and Health Administration; National Institute for Occupational Safety and Health, and the U.S. Department of Energy.

The subcommittee recommends that the NEHC actively communicate the findings and recommendations of HHA reports to a wide array of stakeholders throughout the service, ranging from the client of the HHA to the staffs of the health, safety, and environmental programs. In addition, the basic HHA information should also be made available to all naval personnel and civilian and contract workers, and to the lay public.

Medical Surveillance and Centralization of Medical Data

The Navy's Bureau of Medicine and Surgery collects medical data on Navy personnel. A centralized medical-data management system is being developed for the entire DOD and will eventually include both occupational medicine and industrial hygiene data. Such a system would allow NEHC to access medical data to conduct Navy-wide surveillance studies to detect possible adverse health outcomes related to potential chemical exposures.

The centralized approach for data collection and processing would facilitate communication between various commands within the Navy, such as between NEHC and industrial hygiene personnel at remote locations. Data from the medical-surveillance program should be analyzed by NEHC to evaluate the effectiveness of HHAs in protecting the health of naval personnel on a regular basis. Medical records can be used to verify the effectiveness of the HHA program and its recommendations.

OVERALL SUMMARY

The Industrial Hygiene Directorate's HHA program is designed to protect the health of naval personnel. In a reduced-size Navy, the preventive functions of the NEHC can be an important factor in reducing costs associated with Navy health care and readiness. At present, NEHC's HHAs provide advice only. The subcommittee recommends that the Navy consider elevating the importance of the HHA program (e.g., by delegating authority to the NEHC for "signing off" on decisions to use or not use products) and increasing support for the HHA program throughout the Navy command structure.

The subcommittee concludes that the development of formal, written SOPs; the addition of senior scientists with expertise in toxicology, epidemiology, risk assessment and industrial hygiene; increased training of the current staff; better quality control and quality assurance procedures, including the formation of a peer review board; improvements in data acquisition and management; increased communication between NEHC and other DOD agencies and stakeholders; and the development of a centralized medical-data management system would lead to a more effective HHA process that would stand up to critical and objective scrutiny.

*Review of the U.S. Navy
Environmental Health Center's
Health-Hazard Assessment Process*

1

Introduction

THE NAVY AND MARINE CORPS use a broad range of materials in operations that occur on land at shore facilities, in the air in combat and reconnaissance aircraft, in surface vessels on seas around the world, and in submarine vessels that operate as self-contained environments. Many of these materials are unique to military operations. Some are also used in civilian operations, but the use of these materials by the Navy and Marine Corps can be substantially different. Although many of the materials used by the Navy might be relatively innocuous, there are a large number that can pose significant health hazards under specific exposure circumstances.

The Navy Environmental Health Center (NEHC) located in Norfolk, Virginia, is the primary organization within the Navy that is tasked with assessing occupational and environmental health hazards for naval personnel from exposures to toxic substances. It serves as the central source or corporate center that provides the Navy and Marine Corps, ashore and afloat, with technical support for preventive medicine, medical management, health promotion, drug screening, and occupational and environmental health programs. For many of these programs, NEHC reviews toxicological and related data and prepares health-hazard assessments (HHAs) for potentially hazardous materials under a variety of exposure conditions. Because NEHC is continually being asked to develop HHAs for new materials for the Navy and the

Marine Corps, and because the Navy is committed to protecting its personnel, their families, and communities surrounding the Naval sites from exposures to toxic chemicals, the National Research Council (NRC) was asked to assess the validity and effectiveness of NEHC's HHA process; determine whether the process as implemented provides the Navy with state-of-the-art, comprehensive, and defensible evaluations of toxicological hazards; and identify any program elements that require improvement.

The NRC assigned this project to the Board on Environmental Studies and Toxicology's Committee on Toxicology (COT). The COT convened the Subcommittee on Toxicological Hazard Assessment to address this project. The subcommittee has expertise in general toxicology, inhalation toxicology, epidemiology, neurotoxicology, immunotoxicology, reproductive and developmental toxicology, pharmacology, medicine, risk assessment, and biostatistics.

BACKGROUND

NEHC and its directorates (including subordinate departments) fall under the direct command of the Navy's Bureau of Medicine and Surgery (BUMED). (Appendix A describes the history of NEHC and its relationship with other naval organizations). The Hazardous Materials Department (HMD) of the Industrial Hygiene Directorate is the primary group within NEHC responsible for conducting HHAs of materials or systems that range from single compounds to complex operational systems such as the Tomahawk cruise missile or helicopters. Recent examples of some of the HHAs developed by NEHC pertain to barrier coatings (e.g., antifouling paints and novel "preservative" coatings for surfaces exposed to salt spray), insulating materials, torpedo construction materials, and construction materials for temporary shelters.

HHAs developed for various chemicals used by the Navy play an important role in making decisions regarding the procurement of materials and operating systems by Naval Sea Systems Command—one of the largest acquisition commands of the Navy. HHAs also help to minimize the number and quantity of potentially toxic materials that are integrated into Navy operations by (1) reducing the number of toxic materials bought by the Naval Sea Systems Command and other

Navy commands, and (2) reducing the number of substances or systems integrated into Navy operations that produce hazardous waste. In practice, candidate materials or systems are first screened by the Naval Sea Systems Command or other commands for operational acceptability and cost. Candidate materials or operating systems that meet those criteria are then evaluated for potential human-health hazards by HMD using its HHA process. HMD also conducts life-cycle assessments of materials to minimize generation of hazardous waste. Program managers in the acquisition commands weigh the operational, economic, and health risk factors to choose an ideal candidate material for the Navy. In this way, selections tend to be biased toward a "front-end" reduction of hazardous materials during the procurement process.

HHAs also provide commanders and commanding officers with technical assistance for evaluating and monitoring hazardous materials in the workplace and by recommending precautionary measures, including the development of lists of hazardous materials authorized for use.

Over the past few years, the need and urgency for HHAs has increased dramatically. This is in large part due to the fact that the reduction of crew sizes and the elimination of redundancy among many occupations in the Navy have put emphasis on quality-of-life concerns for naval personnel who are required to perform onerous functions (e.g., chipping paint below decks when in port). In addition, there is increasing reliance on advanced technologies to compensate for reductions in force levels and personnel. These new technologies often require careful analysis to identify their potential to adversely affect health and readiness.

Because NEHC must meet these obligations without any increase in resources, a scientifically sound and effective HHA process is needed. The subcommittee's report is intended to provide NEHC with recommendations that will improve and strengthen the HHA process and the Navy's efforts in preventive medicine.

SUBCOMMITTEE'S APPROACH TO THE CHARGE

The subcommittee's conclusions and recommendations, as presented in this report, are based on its review of documents submitted by NEHC; presentations made by NEHC personnel at subcommittee meetings; and site visits to NEHC and the aircraft carrier, U.S.S. *Constellation*, while docked at the Naval Air Station North Island, San Diego, California. In addition, the subcommittee reviewed the HHA processes used by chemical and pharmaceutical companies for their adaptability and usefulness to the military situation.

STRUCTURE OF THIS REPORT

The remainder of this report is organized into three chapters. In Chapter 2, the subcommittee reviews NEHC's current HHA process. The HHA process used by the chemical and pharmaceutical industry and its applicability and usefulness to the military situation is discussed in Chapter 3. Chapter 4 contains the subcommittee's conclusions and recommendations. Appendix A describes the relationship of NEHC with other Navy organizations; it also describes the role of BUMED with regard to the HHA program. Appendix B presents the policies and instructions issued by Department of Defense and the Department of the Navy with regard to the use of hazardous materials.

2

The Navy's Current Health-Hazard Assessment Process

THIS CHAPTER REVIEWS the Navy's policies, directives, and regulations for handling hazardous materials to determine if the Navy Environmental Health Center (NEHC) is adequately constituted to carry out its mission. This is followed by a review of the Navy's current health-hazard assessment (HHA) process as implemented by NEHC, which includes a discussion of the types of HHAs conducted by NEHC, and the resources, information sources, and quality-control procedures employed.

NAVY POLICIES AND DIRECTIVES RELATED TO HAZARDOUS SUBSTANCES

Because the Navy is a very large organization spread out over the entire marine geography of the earth, it requires an integrated command structure to carry out its duties for sound and responsible handling of hazardous substances. Without a responsive command structure that acts in an effective, efficient, and coordinated manner to handle hazardous materials, the Navy's mission to protect and defend the nation and fulfill treaty agreements with allies could be jeopardized through impaired health of key combat and support personnel, loss of public confidence in its ability to operate effectively, or loss of the good will of those countries that host Navy facilities.

The Department of Defense (DOD), the Secretary of the Navy, and the Navy's Bureau of Medicine and Surgery (BUMED) have published many policies, directives, instructions, and military standards that set the framework for control of hazardous substances within the Navy. This section reviews and summarizes those documents to assess whether the Navy's management system for handling hazardous materials clearly states its policies with regard to how such materials should be handled, who is responsible for implementing the policy decisions, and whether NEHC is adequately constituted to carry out its assigned mission.

A summary of pertinent policies, directives, instructions, and standards, is presented in Appendix B. DOD and Navy documents clearly set out expectations that Navy managers, whether officers, enlisted, or civilians, are responsible for designing and implementing hazardous materials and that they use procedures that are based on pollution prevention (particularly source reduction) and life-cycle review considerations. The relevant policy statements make it clear that there is an expectation that hazardous chemical materials be fully evaluated prior to use, that the lowest hazard material (subject to consideration of operational acceptability) be selected, that safety is built into the design of systems, and that problem prevention rather than remediation is a significant design and use consideration. The elements of Navy-wide health, safety, and environmental programs that are outlined in those documents include (1) life-cycle assessment, emphasizing pollution prevention and programs to acquire less hazardous materials, and (2) hazardous-material control (including hazard identification and risk assessment) by incorporating elements of occupational safety, industrial hygiene, occupational medicine, and hazard communication.

The documents also demonstrate that NEHC is given considerable responsibility for reviewing and assessing the impact of potential hazardous substances on the health of Navy personnel and communicating its findings to the Navy command structure. NEHC is instructed to provide (1) toxicity and related data, (2) guidance on the control of hazardous substances, and (3) recommendations for exposure limits. NEHC is not given responsibility for reviewing or assessing the environmental aspects of hazardous substances, except on an as-requested basis or as a part of the Defense Environmental Restoration Act or Base Realignment and Closure programs; those requests come from the Naval Facilities Engineering Command. NEHC is also not given the responsibility for reviewing and assessing the intrinsic safety properties of materials for fire hazard or explosivity. Although the responsibilities of NEHC are clearly stated, it is not given authority to require Navy managers in the client commands to use its services, or authority to prohibit the use of certain hazardous materials.

In addition, the documents indicate that NEHC is the primary source for review of potentially hazardous substances for the Navy. However, Navy operational units may screen materials—for small on-the-shelf types of purchases—through the safety office at site operations and regional occupational health centers. Only materials that are not cleared during the screening process at the unit level or by regional occupational health centers are sent to NEHC. However, NEHC is the only source for providing HHAs to acquisition commands, such as the Naval Sea Systems Command. HHAs are useful to acquisition commands in making decisions for purchasing large amounts of chemicals that would be used throughout the Navy.

Although this process appears to make efficient use of Navy resources at multiple organizational levels, it also provides some challenges for adequate training of operations and regional staff, consistency in hazard evaluation across organizations, and life-cycle assessment (LCA) of hazardous substances across Navy operations. Although DOD and Department of the Navy documents include policy statements requiring the use of LCA, it is not clear that consideration of LCA is actually taken into account in the procedures that the Navy has put into place to handle potentially hazardous materials.

DOD Directive 5000.1 (Defense Acquisition Management Policies and Procedures) outlines a DOD acquisition process (See phases and milestones process in Figure 2-1) that is similar to many of the "phases and gates" processes used in civilian industrial product-development processes. Although the process does require identification of "potential environmental consequences" at Phase 0, it is not clear if health effects other than those listed on a material safety data sheet (MSDS) are considered in the process or if milestone approvals require sign-off by qualified and experienced professionals on the NEHC or BUMED staff for health assessments. Comparable civilian processes often require such sign-offs to assure that risk assessments are conducted by

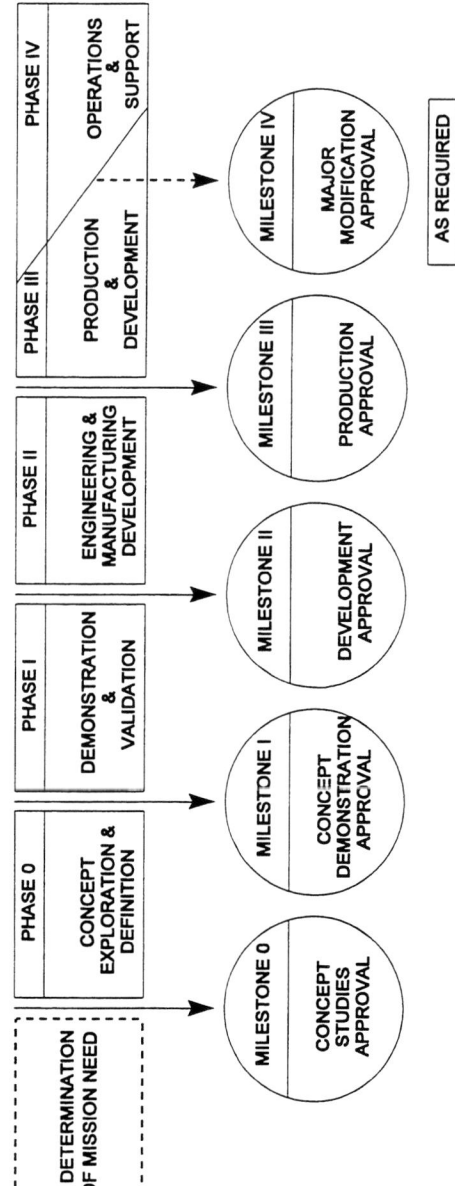

FIGURE 2-1 Acquisition phases and milestones.

qualified personnel who can act independently from the command structure controlling acquisition. DOD and Navy policies also require that end-of-life considerations must be taken into account during acquisition. From examination of the model presented in DOD Instruction 5000.1, the policy contained in this instruction is not complete because the last phase of the process (Phase IV) deals with operations and support rather than disposal and recycling (required elements of a complete LCA).

THE NAVY'S CURRENT HEALTH-HAZARD ASSESSMENT PROCESS

The Navy uses large amounts of potentially hazardous materials. Control over use of hazardous materials necessitates a considerable investment in minimizing the number and quantity of hazardous materials purchased and that resulting waste material that might be generated. This effort ranges from major weapons systems procurement to individual off-the-shelf purchases. Navy policy requires, after suitable life-cycle considerations, that the material with the least hazard potential (operationally acceptable) be selected for use. The NEHC health-hazard assessment program provides commanders and commanding officers with technical assistance for evaluating toxicity and other relevant data, identifying appropriate control measures, monitoring the use of hazardous materials in the workplace, and developing authorized hazardous-materials-use lists.

NEHC's HHA process depends heavily on the availability and quality of toxicity and use information as well as on individuals capable of understanding and applying that information in an appropriate manner. The HHA process provides for several levels of review for toxicity and of related data to prepare HHAs for various chemicals used by the Navy. A tiered approach to assessment of health risk has been used, with a large number of nonhealth professionals within individual commands using standardized criteria to screen materials.

Figure 2-2 shows the process flow for conducting HHAs within the Department of the Navy. Initially, the safety office at site operations (ashore or afloat) screens a substance for operational acceptability and economic feasibility and, based on product information (e.g., an MSDS), evaluates whether it is an occupational health hazard (Level I

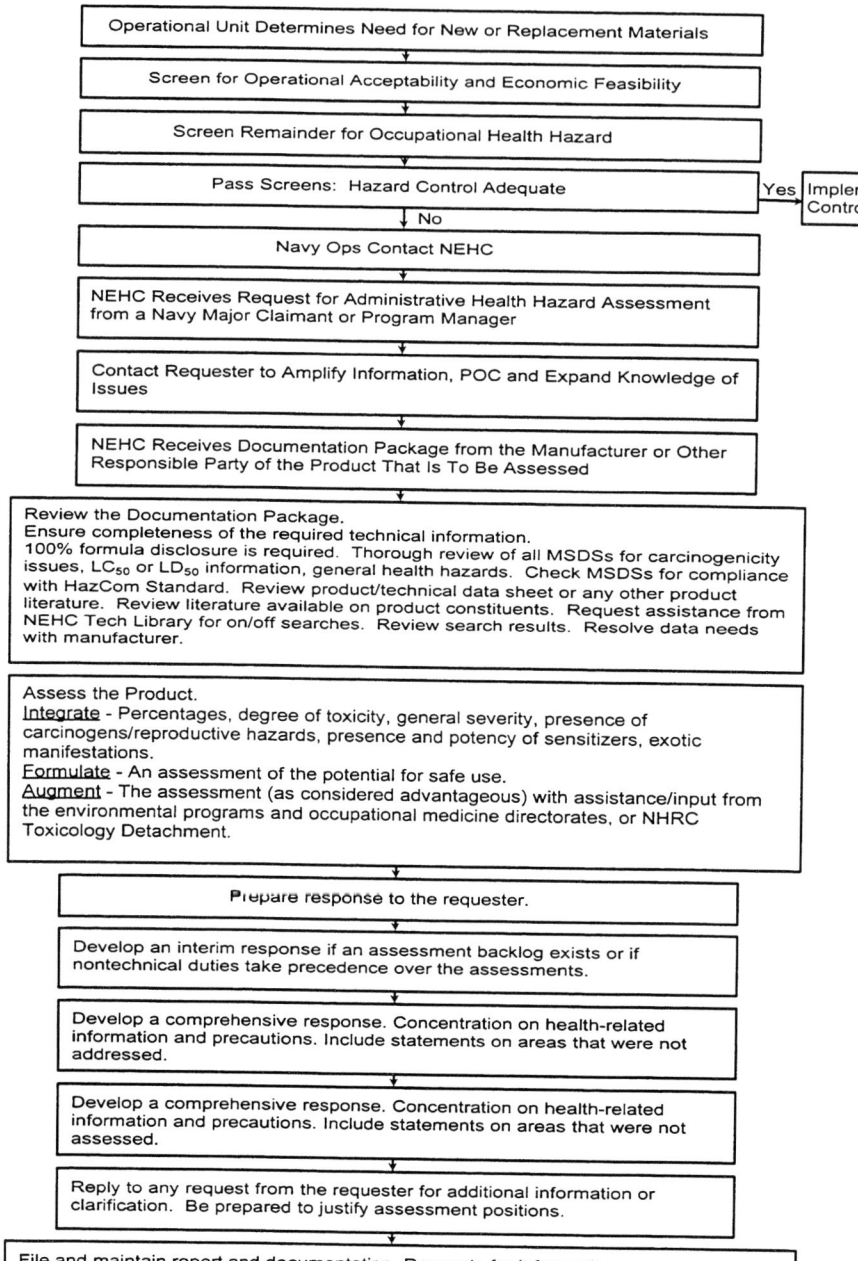

FIGURE 2-2 Process flow chart for health hazard assessments within the Department of the Navy.

review). If the safety office at the site operations cannot determine whether or not the substance is a health hazard, industrial hygienists and occupational medicine physicians at a regional occupational health department then review the material (Level II review) and provide recommendations. If there is still uncertainty about the potential hazards, NEHC is contacted to assess the health-hazard potential (Level III review). If insufficient information exists to complete a Level III review, NEHC might contact the Toxicology Detachment of the Naval Health Research Center (NHRC/TD) to perform a Level IV review which might require toxicological testing and development of quantitative risk assessments. As a matter of DOD and Department of the Navy policy, NEHC uses exposure standards set or recommended by the Occupational Safety and Health Administration (OSHA) and the American Conference of Governmental Industrial Hygienists (ACGIH) for those substances with available values. For Navy-specific substances, exposure standards are developed and recommended by NEHC or the NHRC/TD. The Navy usually asks the National Research Council (NRC) Committee on Toxicology (COT) to review the scientific validity of standards set by NEHC or NHRC/TD. NRC recommendations are reviewed and approved by BUMED and forwarded to the Office of the Chief of Naval Operations for promulgation. In certain circumstances, these might be coordinated across services of the DOD for consistency.

Examples of Health-Hazard Assessments Conducted by NEHC

To evaluate NEHC's HHA process, the subcommittee requested a description of the work of NEHC's Hazardous Materials Department (HMD), which is the responsible body within NEHC that prepares HHAs. In response, a series of documents were received that covered the efforts of NEHC from December 1997 through January 1998. A total of 98 actions were taken. Table 2-1 provides examples of the types of NEHC documents that were submitted to the subcommittee as well as an estimate of the professional level that was thought to be necessary to complete the project successfully. Table 2-2 lists the types of administrative health-hazard assessments prepared between December 1997 and January 1998. Each HHA contains references to the request

TABLE 2-1 Examples of Documents Submitted by the Navy Environmental Health Center and Personnel Required to Perform Specific Projects

Type of Project	Number of Requests/Responses Documented During the Period Specified Below	Personnel Required to Perform Specific Projects
Department of Defense, Hazardous Materials Information System (HMIS). Technical focal point responsibilities. Forwarding of MSDSs.	638 MSDS (10 December 1997 to 30 January 1998)	B.S.-level industrial hygienist
Forwarding of technical documentation packet.	2318 technical documentation packages via electronic media (10 December 1997 to 30 January 1998)	B.S.-level industrial hygienist
Requests from suppliers for information pertaining to the complete chemical content of products undergoing evaluation by the Navy	14 (10 December 1997 to 30 January 1998)	B.S.-level industrial hygienist
Development of HHAs. Requests received from operational command program managers.	74 (calendar year 1998) 81 (calendar year 1997)	B.S.-level/senior B.S.-level/M.S.-level industrial hygienists. Consultation with physicians, toxicologists, environmental protection specialists and preventive medicine specialists as required.
HHAs: Special projects related to complex systems (examples: Advanced Amphibious Assault Vehicle-Fire Suppression System) (Does not include submarine materials reviews that are addressed in Table 2-3)	2 (per calendar year)	Senior B.S.-level/ M.S.-level industrial hygienists. Consultation with physicians, toxicologists, environmental protection specialists, and preventive medicine specialists as required.

for information, the source of the information, indications of whether the material meets screening criteria and whether it is believed that the material can be used safely, a list of precautions under which it

TABLE 2-2 Health-Hazard Assessment Reports Prepared and Issued Between 10 December 1997 and 30 January 1998. (Selected Sample)

Type of Material/Compound or Product Addressed in Health-Hazard Assessments	Number of Health-Hazard Assessments Performed during the Period (Selected Sample) and Command Receiving Assessment
Lubricants and greases	2 (Naval Surface Warfare Center)
Acrylic coating	1 (Naval Surface Warfare Center)
Leak detectors for halocarbons	2 (Naval Surface Warfare Center)
Solventless varnishes and electric motor applications	2 (Naval Surface Warfare Center)
Laboratory chemical for shipboard use	1 Chlorobenzene (Naval Surface Warfare Center)
Tapping and cutting fluids	2 (Naval Surface Warfare Center)
High-temperature polytetrafluoroethylene tape	1 (Naval Surface Warfare Center)
Ion-exchange resin	3 (Naval Surface Warfare Center)
Epoxy resin systems	4 (Naval Surface Warfare Center)
Adhesive film	1 (Naval Surface Warfare Center)
Stain removal materials	2 (Naval Surface Warfare Center)
Battery corrosion preventive compound	1 (Naval Surface Warfare Center)
Life preserver inflation device	1 (Naval Surface Warfare Center)
Metal surface filler	1 (Naval Surface Warfare Center)
Desiccant	1 (Naval Surface Warfare Center)
Powder coatings	8 (Naval Sea Systems Command)
Thermal insulation and coatings/adhesives	4 (Naval Sea Systems Command)
Lubricants	1 (Naval Sea Systems Command)
Cleaners/degreasers	2 (Naval Sea Systems Command)
Epoxy systems and primers	2 (Naval Sea Systems Command)
Industrial finishes	1 (Naval Sea Systems Command)
Paint strippers	1 (Naval Air Warfare Center)
Industrial flooring system	1 (Navy Environmental and Preventive Medicine Unit # 2, Norfolk, VA
Industrial solvents	1 (Naval Sea Systems Command)
Industrial solvents	1 (Naval Facilities Engineering Command)
Industrial solvents	1 (Naval Air Warfare Center)

might be used, first aid instructions (if appropriate), and a contact person. Table 2-3 shows a list of special projects conducted by NEHC.

The majority of projects reviewed did not require additional expertise in toxicology, risk assessment, or biological modeling beyond that currently resident in NEHC. In completing these tasks, NEHC appeared to be placed in the position of a "job shop" in that it appeared that requests for work were sent and reports were promptly prepared and sent out, but there was little or no interaction during the process and no feedback to NEHC on the adequacy of its evaluation.

NEHC'S PROCESS FOR PREPARING HHA REPORTS

The remainder of this chapter reviews the HHA process (Level III review) that is used by NEHC to determine whether the process provides both the Department of Navy and the Marine Corps with state-of-the-art, comprehensive, and defensible HHAs. The subcommittee was not charged with evaluating Levels I, II, and IV reviews or reviewing the approval and the promulgation process of higher authority. The subcommittee's assessment of NEHC's HHA process was based on documents provided by the Navy and presentations made by Navy personnel, observations and evaluations made during a site visit to NEHC in Norfolk, Virginia, and a site visit to the naval aircraft carrier U.S.S. *Constellation* while docked at the Naval Air Station, North Island, in San Diego, California.

Navy programs involved in research, development, testing, and acquisition forward requests for HHAs via their chain of command to NEHC. Those requests are to be made early enough in the developmental phase of each program to allow sufficient time for the assessments to be performed.

The information to be provided to NEHC by the client command includes a description of the material or process under consideration (including composition, physical and chemical characteristics, and other information that might appear on a standard MSDS), description of the intended use of the material, estimates of the numbers and types of personnel who might be required to work with the material, an estimate of the quantities likely to be in use and in storage, information regarding details of a material being replaced, and an outline of the developmental or introduction milestones.

TABLE 2-3 Special Projects Conducted by the Hazardous Materials Department, Including Submarine Material Reviews Prepared and Issued Between 10 December 1997 and 30 January 1998. (Selected Sample)

Project	Type of Project and Comments
Refrigerant 404a use aboard submarines	Provides HHA of refrigerant, notes restrictions for use and assigns a recommended usage category.
Solventless motor varnishes	Provides HHA of these varnishes, provides guidance with regard to varnish use, off-gassing of aldehydes, and assigns recommended usage category.
Washroom cleaner	Provides HHA of washroom cleaner proposed for use on submarines and assigns a recommended usage category.
Rubber mounts and adhesives	Provides HHA of mount/adhesive use and assigns a recommended usage category. Recommends follow-up off-gas testing and further analysis of these materials.
Review/recommendations pertaining to the submarine Materials Test Protocol and Quality-Assurance Program	Provides review of protocol, a listing of target compounds and recommendations for revision and incorporation of new target compounds into the program.
Molybdenum disulfide antiseize compound	Provides review of off-gas testing data from samples analyzed at the National Aeronautics and Space Administration/ White Sands Test Facility (NASA/WSTF), Las Cruces, NM, and assigns a recommended usage category for this antiseize compound.
Armaflex insulation	Provides a submarine materials review and a proposed usage category for this insulation. Recommends follow-up off-gas testing at NASA/WSTF and further analysis of this insulation.
Non-metallic grating	Provides review of grating for use in submarine access tubes. Provides additional guidance regarding hazards associated with installation of grating.
Battery liner coating	Provides review of off-gas data from samples of battery liner coating, which is proposed for use in submarine battery compartments. Recommends usage category.
Enzyme digester	Provides review of detergent. Provides required precautions for use, and recommends follow-up testing of material.
Optical bypass switches	Provides review of optical bypass switches for use in missile control centers. Recommends usage category.

(Continued)

TABLE 2-3 *(Continued)*

Project	Type of Project and Comments
Adhesive for installing rubber sheeting	Provides review of proposed adhesive use, notes potential hazards and control recommendations, and recommends material usage category.
Revision to Nuclear Powered Submarine Atmosphere Control Manual	Provides confirmation that recommendations provided via five previous letters have been included in the final draft of the Manual.
Adhesives	Provides review of all known sources of this adhesive and identifies additional requirements for off-gas testing of materials at the NASA/WSTF. Provides health-hazard control recommendations.
Battery charger	Provides administrative review of battery charger, notes potential hazards and concerns, and recommends a usage category.
Open cell foam panels and fabric covers	Provides review of off-gas testing data for both systems and recommends usage categories. Also recommends pre-baking of systems to minimize subsequent off-gassing when first placed into service.
Development of permissible exposure criteria and health-hazard control guidance for hydroxylammonium nitrate (HAN)	Addresses possible use of HAN as the oxidizer for torpedoes and an internal combustion catapult (new programs development). Discusses possible need for COT participation in development of permissible exposure criteria for HAN.
Powder coatings	Provides HHAs of significant new use of 25 powder coatings. Identifies potential health hazards and control measures. Recommends on-site initial trials prior to final approval for use. Assigns interim submarine material usage categories.

In completing a Level III review, industrial hygiene personnel at NEHC use the scientific literature (information provided in MSDS, toxicology books and journals, electronic literature searches, etc.) and their experience relevant to naval operations to evaluate the available toxicological, epidemiological, and related data; judge the nature and degree of the exposures that might occur; assess the potential health hazard from each; and recommend controls needed in a given use situation to minimize the health risk. If necessary, NEHC also provides interim recommendations to the requesting command to ensure that obvious health hazards identified early in the review process are made known to the users, and provide recommendations for surveillance and control.

When appropriate, other governmental agencies and technical organizations, as well as private consultants, are consulted for their input in preparing HHA reports. NEHC also coordinates with NHRC/TD, located at Wright Patterson Air Force Base near Dayton, Ohio, in performing HHAs. Coordination and interaction between these two groups occurs most frequently when information on the toxicity of the material is limited or when a quantitative risk assessment is required. This interaction generally involves requesting NHRC/TD to review toxicological data, determine possible additional research efforts to fill data gaps, estimate resource availability and project duration, and estimate additional resources that might be required to advance each project to meet the requesting command's deadlines. When additional review is considered necessary, specific questions are then referred to the COT.

In addition, NHRC/TD maintains liaison with other sources of pertinent expertise such as the Air Force is Armstrong Aerospace Medical Research Laboratory and the NRC's COT, performs or obtains needed research to fulfill risk characterization requirements, and keeps NEHC informed of the same. NHRC/TD also maintains toxicology databases that are presumably available for use by NEHC.

Further, staff guidelines for the preparation of HHAs point out that "if it is considered to be advantageous," assistance can be sought from the Environmental Programs Directorate or the Occupational Medicine Directorate for comment on the occupational medicine and environmental hazard aspects of a product's use (J. Drewyer, personal commun. Aug 11, 1997). Hazard assessors are also directed, on an as-needed basis, to request additional consultation from the NHRC/TD according to the procedures outlined in BUMEDINST 6270.8. (See Appendix B.)

NEHC staff state that NHRC/TD provides an extremely high level of toxicological support for the NEHC program on a daily and continuous basis. Assistance contacts are via phone, email, and official correspondence, and occur at least daily, with some "hot issues" requiring hourly contact, and several days to weeks for completion. NEHC staff estimate that at least 500 contacts per year occur, probably requiring more than 1,000 hours of NHRC/TD staff time to research and provide the needed level of support required.

The recommendations contained in HHA reports are incorporated into the requesting command's occupational safety and health program and are used by the requesting command (client) as part of their decision-making to determine whether the development or use of the material should be continued.

An outline of the informal procedures for Level III review by NEHC is summarized below:

- The HHA assessment process is triggered by a request to NEHC for assessment from a Navy operations manager, a program manager, or a safety officer.
- If necessary, NEHC contacts the requesting agency to obtain additional information, points of contact, or clarification of administrative items.
- NEHC receives a documentation package from the manufacturer of the product being assessed, including an MSDS.
- NEHC reviews the manufacturer's documentation package and ensures that the required technical information is complete.
- Essential information for a review includes a complete description of the product, intended use, technical specification sheets and sales literature, supplier's name, supplier's phone number, a technical point of contact, an MSDS that complies with the OSHA hazard-communication standard (HCS), complete product formula with ingredients totaling 100%, Chemical Abstracts Services (CAS) number for each ingredient, a current MSDS for each ingredient, the temperature to which the product will be subjected during use as well as maximum use temperature, copies of toxicity studies related to the product and its ingredients, and copies of standard operating procedures that relate to the application or use of the product.
- Desirable information to complete an HHA includes a small sample of the product as sold, copies of any industrial hygiene survey reports that address potential health hazards related to working with the material, copies of laboratory reports that address the composition and magnitude of pyrolysis products emitted from the product when it is involved in a fire or otherwise severely heated or allowed to contact molten metal, and a technical points of contact within the Navy and at major commercial users of the product should additional information pertaining to application or use experience be needed.

- If information in the documentation package is missing, incomplete, or suspected to be erroneous, NEHC will consult with the point of contact at the manufacturer or the Navy requestor for supplemental information.
- NEHC will review the MSDS for the chemical material and for the constituents of the material, making special note of the presence of known or suspected human carcinogens. These include International Agency for Research or Cancer (ARC) Groups 1, 2A, 2B; the National Toxicology Program (NTP) list of known or suspected carcinogens, and processes that NTP lists as known to be carcinogenic; U.S. Environmental Protection Agency (EPA) assessments on the Integrated Risk Information System (IRIS); or OSHA regulated carcinogens. Previous experience by NEHC staff has demonstrated that a number of noncancer decision criteria serve as useful determinants of toxicity and allow identification of problematic compounds. As a consequence, special consideration is given to chemicals with an oral LD_{50} of less than 500 mg/kg of body weight (rats); an LC_{50} of less than 2,000 ppm by volume of gas or vapor, or 20 mg/liter of mist, fume, or dust; a dermal LD_{50} of less than 1,000 mg/kg of body weight (albino rabbits, 24 continuous hours skin contact); Navy occupational chemical reproductive hazards and sensitizers (dermal, respiratory, and systemic); caustics and corrosives; and highly flammable items and dangerously reactive or explosive materials. The data required by OSHA's HCS on an MSDS are reviewed. Non-HCS data are reviewed for additional information.
- NEHC reviews the manufacturer's product data sheet, technical data sheet, product use sheet, and any other adjunct information that might contribute to better understanding of the materials being assessed.
- NEHC then reviews chemical and toxicological reference materials and computerized databases that are readily available and can provide information on the products or constituents being assessed. Among the most frequently used references are the CCINFO CD-ROM series (Canadian Center for Occupational Health electronic MSDS database), Casarett and Doull's *Toxicology* (Klaassen et al.1996), Micromedex Toxicology, Occupational Medicine and Environmental Services (TOMES), Compendium of Safety Data Sheets for Research and Industrial Chemicals (Keith and Walters 1986), *CRC Handbook of Chemistry and Physics* (Lide 1999), *Hawley's Condensed Chemical Dictionary* (Lewis

1993), *Chemical Hazards of the Workplace* (Proctor et al. 1988), *Handbook of Emergency Chemical Management* (Quigley 1994), *Dorland's Medical Dictionary* (W.B. Saunders Co. 1994), IARC Monographs, NTP Reports, and federal regulations (29 CFR Part 1910 and 42 CFR Part 84).

• As considered necessary, searches are performed by the NEHC library staff. The databases most commonly used by the library staff are CHEMID or CHEMLINE, CANCERLIT (Cancer Literature Database), EMIC (Environmental Mutagenesis Information Center Database), TOXLINE or TOXLIT (Toxicology Literature Information Database), CCRIS (Chemical Carcinogensis Research Information System), IRIS, DART (Developmental and Reproductive Toxicology Bibliography File), and the Chemical Abstracts Registry File.

• NEHC personnel then integrate the information acquired in the review process. Consideration is given to items such as material percentages, the degree of toxicity, the general severity of the hazard (chemical or physical), the presence of carcinogenic or suspected carcinogenic materials, the presence of reproductive hazards, the presence and potency of sensitizers, and the presence of any undesirable or exotic manifestations reported by users of the material being evaluated.

• NEHC formulates an assessment of the potential for safe use of the material for the intended purpose, provided all the precautions listed by the manufacturer are followed explicitly. NEHC then compares the safe-use assessment to the manufacturer's MSDS sheet and resolves any significant differences with the manufacturer.

• As needed, NEHC seeks assistance from the Environmental Programs Directorate or the Occupational Medicine Directorate on the product's environmental and occupational hazards. If additional consultation is warranted, a request is forwarded to the NHRC/TD following procedures outlined in BUMEDINST 6270.8. (See Apendix B.)

• An interim response is prepared if a backlog of assessments exists. This response might indicate that the initial assessment was performed, that the material can be safely used for the intended purpose provided the safety and health provisions in the manufacturer's MSDSs are followed explicitly, and that a comprehensive assessment will follow within a prescribed period of time.

• A comprehensive response is prepared either initially or following the interim response. This response will include the adverse health effects that might be encountered from exposure primarily through the nose, eyes, and skin; the potential for causing cancer in exposed populations; the presence (in any amount) of a Navy occupational chemical reproductive hazard; the first aid actions required as a result of accidents; the personal protective devices necessary to reduce exposure risk; general precautionary statements; training requirements; industrial hygiene and medical department review recommendations; information on what the review did not cover; and information on the procedures for obtaining a submarine-use review. Some reports specifically address certain chemcial stressors of concern (e.g., crystalline silica in nonskid surfaces). Ingestion hazards are not normally addressed due to the unlikelihood of occupational exposure via the ingestion route. Nevertheless, users must be cautioned to review the ingestion hazard statements on the MSDS and be prepared to respond as required by the circumstances.

- As needed, replies to follow-up requests are made for amplification or clarification regarding either the interim or the comprehensive HHA.
- NEHC files and maintains the report and all associated documentation indefinitely in a safe and secure area. Information concerning proprietary elements of an assessment is not divulged to any party without the manufacturer's written permission.
- Requests for information from vendors and other sources besides the client command are addressed by the client command. NEHC does not give HHAs to vendors.

SPECIAL CONSIDERATIONS FOR
RISK ASSESSMENTS: POPULATIONS AT RISK

The Navy is concerned about three populations that might be at risk for adverse health effects from exposure to hazardous materials that are used or encountered in its operations. The first and most restrictive population at risk would include only active military personnel. This at-risk population could be expanded to include family members, especially if there is any possibility that the exposure might be transported outside the occupational setting.

A second population would include civilians employed in various capacities at naval facilities or on ships and who have the potential for indirect or direct exposures to hazardous materials in the course of performing their duties. Depending upon the exposures of concern, this definition could be expanded to include dependents and civilian personnel.

The third and most comprehensive population at risk includes the community residing in the vicinity of the naval base. In addition to the aforementioned populations, this population includes all residents in a specified geographical area with susceptible subgroups such as pregnant women, infants, children, the elderly, and persons with preexisting diseases.

EXPOSURE INFORMATION FOR HEALTH-HAZARD ASSESSMENTS

Many occupational exposures to Navy personnel are similar to those that occur among civilian workers. Exposures are expected to be by inhalation or dermal contact in the vast majority of instances. Oral exposures would be expected to occur only under accidental conditions or with poor personal hygiene.

Unlike many civilian exposure conditions, the Navy must deal with work locations on board ships or aircraft as well as at shore facilities. Particularly on board ship and aircraft, serious considerations have to be given to addressing the inability of crews to avoid accidental exposures in certain circumstances. Therefore, although emergency evacuation of a facility might be seen as a way to control accidental exposures in a civilian workplace, evacuation of ships at sea or aircraft in flight are not options for handling many accidental exposures during Navy operations.

Unlike civilian operations, exposures in the Navy occur under "normal workplace conditions" or under "operational conditions," that is, under exposure conditions driven by military mission performance requirements and under constraints very different from civilian exposure conditions. This subset of occupational exposures encompasses a continuum from training through combat.

Exposures to many materials during normal operations can be similar to that of civilian populations (8 hr/d, 5 d/wk, 50 wk/yr). However, to maintain operational effectiveness, frequent personnel training is conducted. Although exposures during training assignments are of relatively short duration, exposure intensity might be much higher than for civilian personnel using similar materials. The trainers might have routine exposures.

In addition, exposures might be continuous for some Navy personnel. Continuous exposure to low levels of hazardous materials are most likely to occur either in undersea operations, where submarine crews live in a self-contained environment, or during sustained operations in contaminated environments.

Risher et al. (1995) have characterized typical Navy personnel as being a young workforce (a mean age of 27 years) as compared with the civilian workforce (35 years old on average) and generally in better physical condition due to Navy physical- readiness requirements. Under peacetime conditions, Navy tour-of-duty rotations (typically 2 to 3 years) and career development pathways reduce the overall duration of workplace exposure. The shorter tours (2 to 12 months) are almost exclusively training assignments; a "normal" tour is generally about 3 years. Current personnel assignment trends are fostering longer stays in a geographical area, but the rotations among commands are still about 3 years in length. A notable exception is that many enlisted functions aboard ship require the individual to stay with the ship for longer than 3 years—about 4 to 5 years for certain specialties. These factors, however, are not likely to be relevant for civilian employees of the Navy who are expected to have employment characteristics similar to that of the civilian workforce in the private sector.

HUMAN RESOURCES AVAILABLE TO CONDUCT HEALTH-HAZARD ASSESSMENTS

HHAs are conducted in whole or in part by professional and technical staff both the Industrial Hygiene Directorate and Environmental Programs Directorate, which are located at NEHC in Norfolk, Virginia. At the time of the subcommittee's evaluation, those directorates were composed of approximately 37 people, of which 5 were military officers and 32 were civil service personnel (administrative and clerical personnel were not counted). Additionally, the Occupational Medicine Directorate had 5 military personnel (4 officers, 1 enlisted) and 7 civil service personnel who could also provide support for conducting HHAs. NHRC/TD employed approximately 40 persons. NHRC/TD has the capability of conducting experimental studies to determine the toxicological hazard of chemicals used by the Navy and to perform risk assessments for chemicals that involve complex issues.

Staff experience and training in the Industrial Hygiene and Environmental Programs Directorates are predominately in industrial hygiene; however, some staff are trained in environmental engineering, environmental sciences, and chemistry. About one-half of the staff had earned master of science degrees in a relevant field and the rest of the staff had earned bachelor of science degrees in a relevant field of study. Only one staff member in those two directorates has a doctorate-level degree in a relevant scientific discipline. However, several persons at NHRC/TD hold doctorate degrees and are available to provide help to NEHC.

INFORMATION SYSTEMS USED FOR HAZARD ASSESSMENTS

Within NEHC, information searches are conducted primarily by the on-site library staff. The kinds and quality of databases and literature sources in use by NEHC staff for use in performing health risk assessments are primarily those available through the National Library of Medicine (NLM) network of databases and the CAS databases. The first search is the CHEMID database for substance identification. CHEMID provides a list of all NLM databases that contain information on the CAS number or name. Each listed database is then searched individually. These include CHEMID, CANCERLIT, CCRIS, DART, IRIS, MEDLINE, Hazardous Substances Data Bases, TOXLINE, and Registry of Toxic Effects. If clarification or expansion is necessary during the search process, there would be an exchange between the NEHC requestor and the librarian performing the search. This exchange would serve to clarify what was desired and amplify or expand on a provided product, as necessary.

Storage of NEHC completed reports and their distribution to NEHC clients appeared to be largely a manual process, rather than an electronic process. Currently, the Navy does not have a computerized database-management structure in place that would permit the identification, evaluation, and control of health hazards for populations at risk.

PEER REVIEW OF REPORTS

In the preparation of an HHA for a chemical material, NEHC is directed by BUMED (BUMEDINST 6270.8, 4.c. (1), p. 5; 6 Jun 90) to seek review of its HHA reports by credible groups. Those requests for review were the mechanisms employed to obtain external review of the Jinkanpo Incineration Complex draft health risk analysis and the "Human Health Risk Evaluation for Past Firefighters' Activities at Naval Air Station Alameda California."

The Naval Air Station Alameda assessment is highly visible due to potential occupational exposures from PCBs, dioxins, and lead incurred during training exercises by fire fighters. The Jinkanpo assessment addresses a highly visible public health issue involving exposure of Navy staff, dependents, and facility tenants to industrial waste incinerator emissions generated by a privately owned, off-site firm regulated by a host country.

NEHC's criteria for determining which assessments undergo external review, and the degree of external review to which the assessment is to be subjected, are discussed below. NEHC policies and procedures require that all HHAs undergo an internal review. External reviews are sought when the HHA or health risk assessments are performed to resolve a controversial issue for which a review by independent groups such as the National Research Council of the National Academy of Sciences would be beneficial when dealing with low-trust and high-concern situations, such as for the Jinkanpo Incinerator Complex at Atsugi, Japan, and the Alameda Fire-fighter risk assessments. Criteria for significant external review appear to be

- high-visibility subjects involving international differences in the interpretation of regulatory compliance (e.g., Jinkanpo Incineration Complex at Atsugi, Japan);
- regulatory issues involving unique occupational exposure scenarios (e.g., Past Firefighters' Activities at Naval Air Station Alameda);

- the introduction of novel compounds (e.g., hydrofluoroether (HFE)-7100 and hydrofluorocarbons (HFC)-236 fa) in the naval supply system.

Selection criteria for identifying which institutions and staff qualifications are needed to perform a given review are not known to the subcommittee. The routine NEHC criterion is to select external reviewers who have risk-assessment experience and are associated with the governmental regulatory sector. If potential external reviewers are from other than the regulatory sector, they are selected because of the basis of their expertise in the primary issue of concern (e.g., PCBs, dioxins). In the case of Atsugi, NEHC also sought reviews from the NRC's COT and the University of Florida. At present, the subcommittee's impression is that the current review of low-visibility tasks is not systematic and might not be sufficiently supported.

3

Other Related Health-Hazard Assessment Processes

IN THIS CHAPTER, the subcommittee reviews other related health-hazard assessment (HHA) processes currently used by the consumer product and pharmaceutical industries and by the U.S. Army Center for Health Promotion and Preventive Medicine (CHPPM).

CONSUMER PRODUCT AND PHARMACEUTICAL INDUSTRIES

To provide content for critically evaluating the Navy's HHA process, the subcommittee reviewed processes, procedures, and best practices used by the consumer product and pharmaceutical industries for determining the safety of new products and materials. The information contained in this section was derived from material provided by major consumer product and pharmaceutical companies (Lilly Research Laboratories, Wolf (1987), Boylstein et al (unpublished); Eastman Kodak Company (O'Donoghue 1989); and Sanofi Research (Dean, unpublished)). Most companies subscribe to the philosophy that their product shall be perceived to be safe under conditions of intended use and reasonably foreseeable misuse. To achieve that objective, companies develop and implement human and environmental safety-assessment programs to ensure that the new products are safe. The responsibility for safety testing is given to a responsible toxicologist in each company. The detailed information needed by the toxicologist to evaluate the safety of the product or its components and to prepare the safety assessment includes the following:

- What is the chemical composition of the product?
- How is the product manufactured?
- Are there known impurities or additives present in the finished product?
- What are the chemical and physical characteristics of the product?
- What will be the conditions of use and exposure?
- What is the duration and route of exposure?
- How much active material will be used in the product?
- How much of each of the component or components can be leached or extracted from the product under consumer-relevant conditions?
- How will the component materials be incorporated into the product?
- Are there analytical data that could be potentially useful for assessing the expected extent of exposure?
- What is the extent of available toxicity literature on the product or class?

Once this information is obtained and the necessary animal tests are performed, the toxicologist then conducts the human-safety assessment. This assessment involves consideration of the toxicological properties of a material and the degree of expected exposure to humans. Exposure assessment requires a realistic balance of information characterizing concentrations, route, uptake, frequency, and duration. Although the details might vary, the basic assessment process is the same for either products consumed by or applied to humans. A typical decision tree is shown (Figure 3-1). It provides the responsible toxicologist or industrial hygienist conducting the risk assessment with a system for evaluating available toxicity information and identifying data gaps.

Once the health risk assessment report is completed, it goes to a report coordinator who identifies internal or external primary reviewers and coordinates the review and sign off. Usually 2 to 4 weeks are allocated for review. Any request for additional reviewers or consultations with scientific experts (secondary reviewers) is the responsibility of the primary reviewer. In most organizations, the report is shared electronically during the review process. A comment memorandum prepared by the reviewer goes to the report writer, study toxicologist, and report coordinator. Once the comments of reviewers have been incorporated into the revised report, it goes to a Chemical Safety Evaluation and Exposure Limits Committee composed of upper-level management with expertise in toxicology, industrial hygiene, environmental affairs, occupational medicine (often, a physician), pharmacology (pharmaceutical industry only), project management, and law (not at all companies). Once the exposure assessment and exposure limits are accepted by the safety committee, the information is directed to management at all sites where the material is utilized or manufactured. It should also be noted that, as additional information becomes available, the risk assessment is periodically reviewed and revised as needed. The universal challenges to risk-assessment groups in industry are to

- provide data for safe workplace utilization or consumer consumption;
- do it quickly and in a cost-effective manner;
- extrapolate from oral doses in animals or humans to airborne concentrations safe for workers in the absence of inhalation data; and
- manage requests for risk assessments with reasonable timeliness.

Most industrial groups maintain an electronic database of information on all their products, intermediates, and additives. That information is updated on a regular basis and available electronically by read-access at all sites globally. The risk-assessment specialist takes advantage of available electronic databases to obtain the best information possible for conducting health risk assessments. Because most of the products are novel, extensive animal tests are performed by the companies either in-house or in a contract laboratory. Even if the studies are performed in contract laboratories, sufficient expertise is required by the company to develop protocols, monitor studies, and evaluate the data obtained.

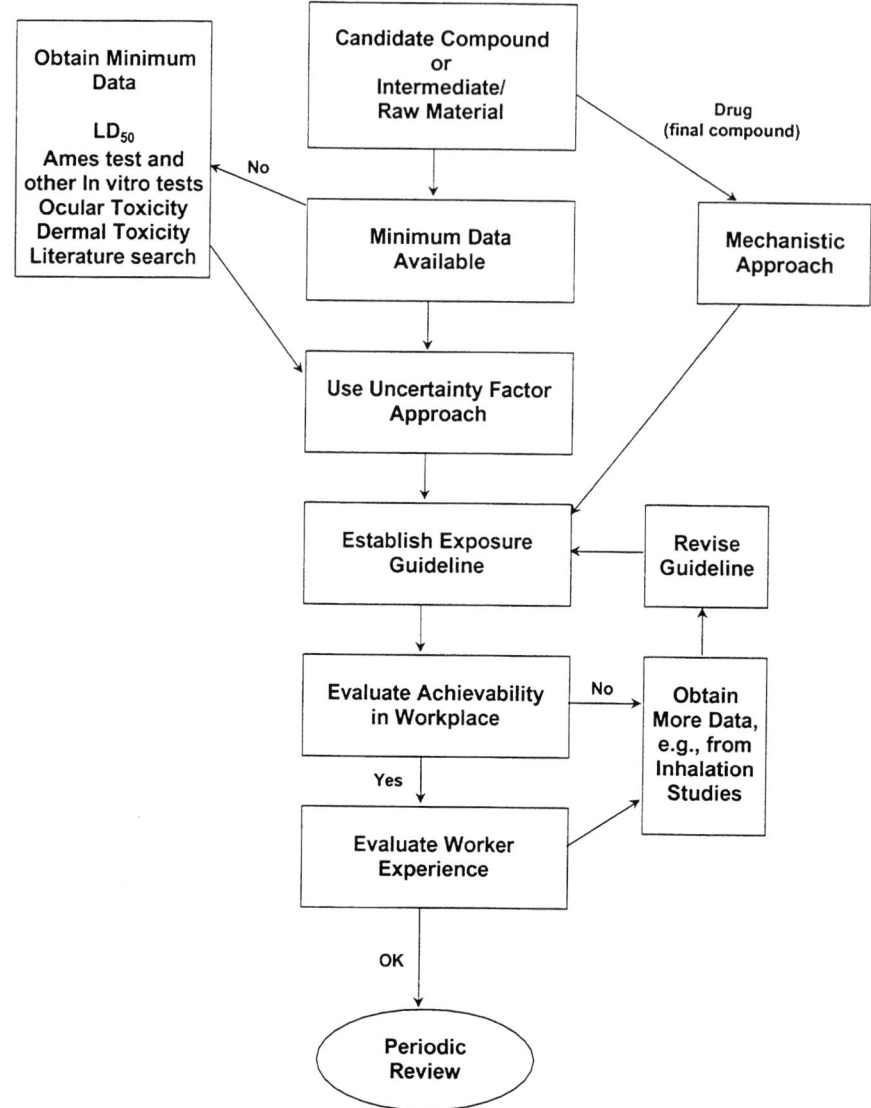

FIGURE 3-1 Typical decision tree for performing toxicological risk assessment (figure provided by R.K. Wolff, Lilly Research Laboratories).

Many companies also utilize closed-loop hazard-assessment processes such as that illustrated in Figure 3-2. In such processes, industrial hygiene data and medical surveillance information are fed back

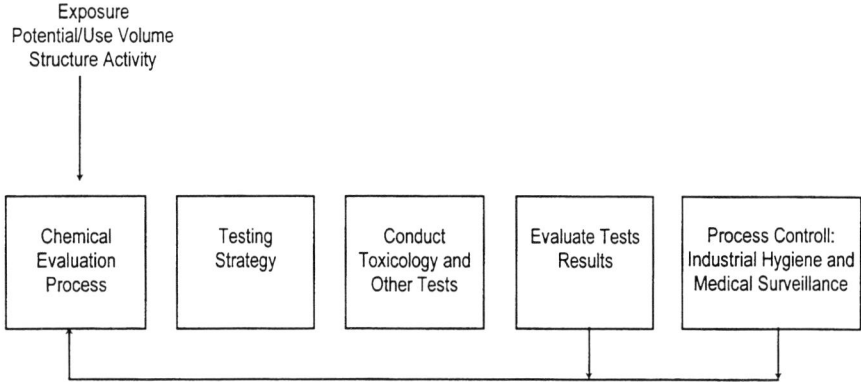

FIGURE 3-2 An example of a closed-loop hazard evaluation process. The chemical evaluation process includes collection of available information, development of a testing strategy, implementation of that strategy, and feedback to the evaluators based on industrial hygiene and medical surveillance information (modified from O'Donoghue 1989).

into the health hazard assessment process to verify the original decisions made on protective equipment and to aid in designing safety standards for new processes (O'Donoghue 1989).

U.S. ARMY CENTER FOR HEALTH PROMOTION AND PREVENTIVE MEDICINE

The subcommittee observed that there are a number of parallels between the Navy Environmental Health Center's (NEHC's) HHA process and that for the U.S. Army Center for Health Promotion and Preventive Medicine (CHPPM) (formerly the U.S. Army Environmental Hygiene Agency), which was the subject of a previous National Research Council review (NRC 1991).

In response to recommendations provided in the review (NRC 1991), CHPPM established a Peer Review Board on Toxicology in January 1992 and created the Quality Assurance Office (QAO) at the command level. CHPPM officials believe that the Peer Review Board has greatly improved the quality of the work being conducted in the organization.

Creating the QAO as a part of the command was a deliberate decision made to ensure independence from personnel engaged in the direction and conduct of toxicological studies. At the same time, the position of Quality Assurance Coordinator was also established.

Standard operating procedures (SOPs) were also developed and communicated to all pertinent staff, a systematic review of the SOP's was implemented, and third-party audits of the QAO and toxicology laboratory for good laboratory practice (GLP) compliance were initiated and maintained. Peer review of test protocols is now routine, and independent review of procedures is periodically sought. The Peer Review Board meets twice yearly to examine progress on outstanding issues, review current and planned work, provide guidance on breaking issues in the field, suggest problem resolutions, and identify useful resources and procedures for data collection and evaluation. Membership is comprised primarily of doctorate-level toxicologists with expertise in industrial toxicology, forensic toxicology, clinical toxicology, pathology, quantitative risk assessment, emergency-response planning, risk communication, analytical chemistry, preventive medicine, toxicological information storage and retrieval systems, and industrial chemical registration and preparation of health-hazard assessments under OSHA and international workplace standards.

The quality assurance coordinator has broad responsibilities to ensure that toxicological studies are performed in compliance with the GLPs, Animal Welfare Act requirements, accreditation criteria of the American Association for the Accreditation of Laboratory Animal Care, and other documented standards. The QAO's primary responsibilities are to

- monitor toxicological studies to assure the management that the facilities,
equipment, personnel, methods, practices, records, and controls comply with the GLP monitored studies include those performed at satellite and analytical contract facilities off-site;
- assist in developing a Facility Master Schedule Database, and maintaining a copy of that for all studies;
- review and maintain SOPs for all toxicological studies and equipment as well as SOPs for inspection functions and responsibilities;
- assist in, and conduct, audits to maintain International Standard-

ization Organization (ISO) compliance and assure follow-up on recommended corrective actions;
- assist in the development of award contracts for services and ensure their compliance with GLP and ISO;
- act as liaison for external Food and Drug Administration or U.S. Environmental Protection Agency inspections;
- maintain copies of all protocols pertaining to toxicological studies for which the QAO is responsible;
- inspect toxicological studies at intervals to ensure study integrity;
- maintain records of inspections and report their results to management;
- review final study reports to ensure that reported results are accurately reflected; and
- maintain SOPs describing the functions and responsibilities of the QAO.

The QAO coordinator periodically reports to the Peer Review Board on the status of GLP compliance in the CHPPM Directorate of Toxicology. The Peer Review Board advises the QAO on possible problem resolution.

The Directorate of Toxicology also maintains its own GLP coordinator, who has responsibility for maintaining Directorate SOPs, communicating SOP requirements to investigators (via both library hard-copy and read-only electronic versions), and implementing corrective actions to resolve QAO audit findings when such actions involve SOP changes or other documentation. As SOP custodian, this individual also reviews new SOPs, maintains current SOPs, and ensures that Directorate personnel who are proficient in performing the SOP procedure review the pertinent SOPs annually for accuracy. This coordinator also ensures that all changes to SOPs or quality documentation are understood by appropriate personnel and develops and implements archiving systems. The QAO coordinator meets often with the Directorate GLP coordinator to reach joint resolution on various issues.

Currently, the Directorate of Toxicology performs health-effects research and toxicity evaluation. Health-effects research staff evaluate methods for predicting and assessing the effects of military environmental contaminants on human health and the environment, and act as consultants to installations and Department of Defense components on environmental and ecological risk evaluations and assessments. In this capacity, CHPPM and NEHC staff have collaborated in the development and presentation of risk-communication workshops at a number of military facilities. Toxicity-evaluation staff within the Directorate identify chemical hazards and perform toxicity evaluations to develop preventive procedures for avoiding or minimizing hazardous exposures. Computerized databases, literature surveys, laboratory studies, and consultations with other health care advisors are all employed to accomplish individual assessments.

The successful results of the QAO and the Directorate GLP coordinator have led to the recent development of an overall quality plan, for the purpose of ensuring that quality assurance "responsibilities, policies and procedures . . . are identified, documented and consistently followed by each organizational element to include [the] continental United States, Subordinate Commands and CHPPM-Europe and Japan" (CHPPM 1999). This regulation also includes an appendix outlining the minimal essential elements of a quality plan. Although the CHPPM quality plan is "applicable to all USACHPPM operations, activities, and contractual services and personnel," it should be considered only as a benchmark example and not a template for what is needed within the Navy.

The Committee on Toxicology

The National Research Council Committee on Toxicology has published several reports on methods to derive exposure guideline levels for short-term and continuous exposures (NRC 1986a, 1992, 1993, 2000a). The subcommittee recommends that NEHC routinely review those reports to help it prepare HHAs for exposure of military personnel and civilian workers to hazardous chemicals.

RECOMMENDATION

The subcommittee recommends that NEHC review HHA processes utilized by industrial and pharmaceutical companies, CHPPM, and COT in preparing its HHAs.

4

Conclusions and Recommendations

THE SUBCOMMITTEE was charged with assessing the validity and effectiveness of the Navy Environmental Health Center's (NEHC's) health-hazard assessment (HHA) process and determining whether the process as implemented provides the Navy with state-of-the-art, comprehensive, and defensible evaluations. The subcommittee was also asked to identify any program elements that require improvement.

The subcommittee's assessment of NEHC's HHA process is based on its review of documents submitted by NEHC, presentations made by NEHC personnel at subcommittee meetings, and site visits to NEHC in Norfolk, Virginia, and the aircraft carrier U.S.S. *Constellation*, while docked at the Naval Air Station North Island in San Diego, California. In addition, the subcommittee reviewed HHA processes used by chemical and pharmaceutical companies for their adaptability and usefulness to the Navy situation.

Based on its assessment, the subcommittee concludes that NEHC's HHA process is adequate for preparing routine HHAs, considering the available resources. There are several deficiencies, however, especially for conducting complex HHAs. To address those deficiencies, the following are needed: (1) formal, written, standard operating procedures (SOPs) for preparing HHAs, (2) staff with expertise in toxicology, epidemiology, and risk assessment for preparing complex HHAs, (3) access to electronic databases, (4) enhanced quality-assurance and quality-control procedures, (5) increased coordination and information transfer between NEHC and other organizations and stakeholders, (6) enhanced medical surveillance and centralization of medical data, and (7) life-cycle assessments. The subcommittee's conclusions and recommendations with respect to each of these major deficiencies are discussed below.

DOCUMENTATION AND DEVELOPMENT OF STANDARD OPERATING PROCEDURES

In reviewing NEHC's procedures for Level I and II reviews, it became apparent to the subcommittee that no formal, documented procedures (e.g., SOPs, including flow charts) have been developed as to how Level I and II reviews should be conducted and documented. In addition, the criteria used to defer reviews to a Level III or Level IV review were not developed. Although this ad hoc process provides a great deal of flexibility which may be important (e.g., when ships are away from home port), it presents challenges for: (a) developing a uniform standard for evaluating health hazards among Navy industrial hygienists, (b) having a Navy-wide hazardous substances control and reduction process, and (c) assuring continuity of NEHC's occupational safety and health programs following the rotation of current industrial hygiene personnel to different duty assignments. The subcommittee believes that the efficiency, consistency, and "institutional memory" of NEHC can be enhanced if formal, documented SOPs for preparing HHAs are developed and implemented. Therefore, the subcommittee recommends that NEHC develop (and update regularly) a set of SOPs for the preparation of its HHAs by incorporating the relevant aspects of procedures employed by pharmaceutical and chemical companies, governmental organizations (e.g., U.S. Army Center for Health Promotion and Preventive Medicine [CHPPM], the U.S. Environmental Protection Agency [EPA]), and the National Research Council [NRC]). The subcommittee also recommends that NEHC develop guidelines or criteria for developing HHAs for use by industrial hygiene personnel on ships or at regional naval occupational health departments. The subcommittee recommends that criteria and guidelines be developed for deciding when to defer a review to NEHC.

Conclusions and Recommendations 47

STAFFING

The effectiveness of NEHC's HHA program is directly dependent on the training and expertise of personnel tasked to prepare the HHAs. For this purpose, the Navy uses experienced professionals who are familiar with Navy operations to provide opinions about potential exposures to hazardous substances.

The professional staff of NEHC's Industrial Hygiene Directorate has extensive experience at the practical level, ranging from 4 to 30 years, in all aspects of hazardous materials evaluation and in the preparation of comprehensive HHAs of new materials and operations contemplated by the Navy prior to introduction. Thus, NEHC has a unique national resource in its staff of 37 professionals who have in excess of 500 years of combined experience in industrial hygiene and environmental health, and in preparing HHAs.

The impression gained from the site visit to NEHC in Norfolk, Virginia, by the subcommittee was that the NEHC staff, from the highest level down, is enthusiastic, hard-working, and intent on making the best possible effort to provide the Navy and Marine Corps with state-of-the-art evaluations of toxicity and potential hazards.

The subcommittee believes that the current education and experience level of NEHC staff are adequate for preparing routine HHAs. This conclusion is based on the subcommittee's evaluation of HHAs and other work submitted by NEHC to the subcommittee for review. Much of the work performed by NEHC can be carried out by scientists or industrial hygienists at the bachelor of science level. In some cases, there is a need for personnel with expertise in industrial hygiene at the master of science level. The subcommittee believes that the current senior civilian staff members are well equipped to deal with many of the routine industrial-hygiene problems. However, there were a few complex risk-assessment projects that required personnel who are highly trained in toxicology, epidemiology, human-health risk assessment, and atmospheric modeling. Examples of such complex projects include those that involve determining the health hazards associated with the off-gassing of chemicals in submarine environments and the risk to personnel and their families from exposure to emissions from the Japanese incinerator in Atsugi. The subcommittee recommends that NEHC recruit professional scientists with expertise in toxicology, epidemiology, and risk assessment for conducting such complex tasks. The subcommittee also recommends that particular attention be given to the qualifications necessary for those personnel exercising key technical oversight and review functions for HHAs.

Industrial Hygiene Personnel

The Navy's worldwide occupational health program is staffed by professionals in a number of fields. Among these, the Navy's industrial hygiene officers are critical to maintaining the health and safety of all naval personnel and contractors who live and work on board ships and are deployed worldwide in support of operational contingencies. They have important roles in prevention, operational risk management, and emergency response.

During a site visit to the aircraft carrier U.S.S. *Constellation*, the subcommittee observed that there was only one industrial hygiene officer and no other personnel in the industrial hygiene unit. There were more than 5,000 personnel on that ship performing a variety of industrial and other procedures that included working with a myriad of chemicals and chemical mixtures, as well as with fuels and munitions. It is not possible for one industrial hygienist to ensure that both work practices and the handling of chemicals in such a complex and highly variable environment are performed in a manner that would ensure the least possible risk to the health of all personnel. The work of the local or regional industrial hygienists can be enhanced, however, with more direct access to NEHC. If a formal relationship between NEHC and shipboard industrial hygiene personnel were strengthened, industrial hygiene officers might be in a better position to foster a safer work environment on ships. One practical approach would be to establish direct computer links between ships and NEHC.

However, because all naval operations need to have an adequate level of access to industrial hygienists, commensurate with the size of the facility and the mission (e.g., an aircraft carrier would need several shipboard billets, whereas a submarine would only need periodic access to such officers), the subcommittee recommends that NEHC evaluate whether Navy facilities have sufficient level of support in industrial hygiene. Large facilities, such as aircraft carriers, should have more than one industrial hygienist to support continuous or sustained operations typical of deploying naval vessels, and in case one of the officers becomes ill, injured, or transferred. In addition, given the unique features of each facility requiring industrial hygiene and hazardous materials/waste officers, there is a need for ongoing training to enable smooth transition when officers are rotated and to provide immediate back up when the primary industrial hygiene officer is not available. Using a "hot fill" approach does not provide the needed expertise for these types of essential operations. The training or cross-training of medical or other officers in basic industrial hygiene and hazardous materials management should be considered.

Another factor necessary to elevate the importance of the industrial hygiene program is to provide NEHC with authority/accountability for mission success. At present, NEHC's HHAs provide advice only. However, delegating authority from BUMED to NEHC for "signing off" on decisions with health risk implications for product use would make any override decision by users (e.g., Naval Sea Systems Command) more highly visible and accountable.

Operations with Reduced Personnel

Because of budget reductions in the Navy, there are decreased numbers of experienced personnel; however, there is an increased demand for greater numbers of HHAs. The combination of decreased personnel and increased demand for HHAs requires a more effective approach for developing HHAs. The subcommittee recommends that NEHC develop a long-term strategy to deal with increasing demand for services in the face of declining resources. This strategy would include elements such as streamlined processes to conserve staff time; increased training of current staff to keep current with advances in toxicology, epidemiology, and risk assessment; and a workforce-planning strategy to replace retiring or relocated staff to ensure the basic quality and quantity of HHAs. This should include a succession plan for NEHC staff, and a projection of future personnel needs.

To streamline the HHA process, the subcommittee recommends that NEHC consider using different levels of review for different levels of potential hazard. In addition, clients should inform NEHC in advance that they will be submitting requests for HHAs. That will allow NEHC time to plan and assemble the necessary expertise to conduct a quality evaluation.

The subcommittee also recommends that NEHC consider task flow rates to determine whether the number and experience of available staff are sufficient for the workload. If not (and there will be peak activity, regardless of how well the program is managed), the subcommittee recommends that NEHC employ outside contractors (e.g., qualified consultants or other military personnel from Department of Defense's [DOD] U.S. Army Medical Research Detachment at Walter Reed Army Institute of Research, USAF Institute for Environmental, Safety, and Occupational Health Risk Analysis [IERA], the Navy Toxicology Detachment at Wright-Patterson AFB, or CHPPM) to maintain quality. It might also be feasible to ask large-procurement bidders (manufacturers or suppliers) to provide HHAs (not just MSDS) as part of their bids.

Training

Training is often the first activity to be reduced in response to budget cuts. Although this would allow more staff time to be devoted directly to the development of HHAs, the subcommittee recommends that specific directives be issued by the Navy to ensure continued HHA-development training for NEHC professionals. Areas of training should include industrial hygiene, toxicology, epidemiology, chemistry, and risk assessment. In addition, the subcommittee recommends that such training be outsourced rather than provided in-house. That is likely to be more cost-effective and productive. In addition, NEHC should make resources available to encourage staff professional development through participation in scientific societies and meetings.

DATA ACQUISITION AND MANAGEMENT

The subcommittee observed that there is an absence, of or limited availability of, computerized hardware for use by key personnel. This impairs access to information for the identification of hazards or human health risks via the internet, intranet, and email and continues to foster reliance on paper records. The absence of a data-management structure precludes meaningful analysis of the vast array of existing data available throughout the Navy on chemicals, exposures, and health outcomes.

During its site visit on board the U.S.S. *Constellation*, the subcommittee had an opportunity to review information systems available aboard ship. In general, there were a limited number of personal computers that are capable of accessing health-hazard information. Material safety data sheets (MSDSs) were provided to the ship industrial hygienist on CD-ROMs. These were updated periodically. However, there were only a few copies of the CD-ROMs on shipboard and access to them was limited. Work areas received MSDSs in paper form after being printed from the CD-ROM. In all the shipboard work areas visited, paper copies of the MSDSs were readily available to the crew and the crew members appeared knowledgeable about their use and content.

The efficiency of the current system for managing MSDSs is of concern. At present, MSDSs are transferred from NEHC for publication by the DOD on CD-ROM. Ganak (1998) has reported that the process for obtaining MSDSs from NEHC through DOD to Navy sites has not been working well and that a solution is being implemented. According to Ganak, it is not unusual for 8 months to elapse before an MSDS appears on a CD-ROM. Because the use of the MSDS has become a critical element of the HHA process, the subcommittee recommends that industrial hygiene personnel at regional or local occupational offices as well as those aboard Navy ships be provided with MSDSs as soon as possible. However, the subcommittee recommends that ship IH officers not solely rely on MSDSs because they sometimes over-estimate effects or fail to adequately deal with serious clinical effects.

The subcommittee believes that high-efficiency information searches are essential to conserve the valuable time of the hazard assessor and recommends that up-to-date computer hardware and software be provided to NEHC's industrial hygiene staff for electronic searches. The subcommittee also recommends that training for the use of that technology be provided.

INFORMATION SOURCES FOR CONDUCTING HEALTH-HAZARD ASSESSMENTS

There are a number of highly credible information sources on the toxicology of industrial and commercial compounds that could provide valuable technical input in developing HHAs. For example, the subcommittee believes that the NEHC will find it useful and cost-effective to more directly and routinely use the EPA's Integrated Risk Information System [IRIS] as screening tools. IRIS is now available at no cost on the Internet. However, IRIS and other secondary references need to be used prudently. For example, if the IRIS entry was completed recently, the information is relatively up-to-date. However, if it was completed several years ago, a re-evaluation of the database may be warranted. NEHC needs to implement additional ways to maintain the currency of toxicological databases (particularly in the time intervals between receiving vendor updates) and to systematically verify (at least a sample of) input data from its current toxicological assessment software.

NEHC now relies heavily on the SmartRisk/SmartTox assessment software to develop some of its HHAs. The SmartRisk/SmartTox software appears to be a good guide for the development of HHAs. However, the information it contains might not always be current and must be updated regularly; the subcommittee recommends that updates be performed routinely. EPA's *Exposure Factors Handbook* (EPA 1997) can be consulted for updated exposure-factors information.

The subcommittee recommends that NEHC not rely heavily on the toxicity information contained within MSDSs in conducting HHAs. The subcommittee believes it is appropriate to begin an assessment with consideration of the MSDS, but all data contained in them should be independently confirmed before use. MSDSs often contain incomplete toxicity information and do not contain the most up-to-date toxicity data available. A cross-check of manufacturer- or vendor-supplied MSDS data for completeness and accuracy is imperative. Information sources available for this purpose include the Hazardous Substances Data Base, Registry of Toxic Effects, the *Agrochemicals Handbook* (Royal Society of Chemistry 1994), *Sax's Dangerous Properties of Industrial Materials* (Lewis 1996), *Patty's Industrial Hygiene and Toxicology* (Clayton and Clayton 1993), the International Agency for Research on Cancer series, and the Air Force toxicology guide (*Installation Restoration Program Toxicology Guide*, ORNL, 1989, 1990). A number of these sources are already in use by NEHC staff.

The subcommittee recommends that NEHC independently confirm key information presented in HSDB, RTECs, and IARC documents.

To avoid duplication, the subcommittee recommends that NEHC thoroughly evaluate existing information from other governmental agencies, other organizations, and the open literature prior to conducting its own HHAs. Cooperative information exchange arrangements should be developed with those agencies and organizations. The Air Force *Installation Restoration Program Toxicology Guides* are an excellent source of toxicity information for chemicals used by the military; they also contain handling precautions. Although these guides were originally designed to address contaminants in drinking water, they also address inhalation and dermal exposures and environmental fate. They cover many chemical compounds that are likely to be used by the Navy. Similarly, the Defense Technical Information Center (DTIC) database is an additional source of toxicity information for chemicals that are of interest to the military.

The subcommittee also recommends that NEHC consider using acute exposure guideline levels (AEGLs) in the preparation of HHAs as an objective source of inhalation toxicology data. AEGLs for a number of hazardous chemicals have been developed by the National Advisory Committee on Acute Exposure Guideline Levels for Hazardous Substances. The AEGL documents are also reviewed by NRC's Committee on Toxicology (COT). Spacecraft maximum allowable concentrations and emergency exposure guidance levels developed by COT should also be used or considered in developing HHAs (NRC 1984 a,b,c, 1985 a, b, 1986 b, 1987, 1988, 1994, 1996 a, b, 2000 b). The subcommittee also recommends that NEHC consider using exposure limits recommended by other governmental and nongovernmental organizations, such as those recommended by the Occupational Safety and Health Administration, National Institute for Occupational Safety and Health, American Conference of Governmental Industrial Hygienists, and American Industrial Hygiene Association. A number of these sources are already in use by the NEHC Staff.

QUALITY ASSURANCE AND QUALITY CONTROL

NEHC's procedures for awarding contracts were not clear to the subcommittee. The subcommittee recommends that NEHC establish a quality-assurance/quality-control (QA/QC) program to assist in the development of contracts for services and establish a means for ensuring the quality of such services.

NEHC's selection criteria for the identification of institutions and staff necessary for the performance of internal or external reviews were also not completely evident to the subcommittee. It is the subcommittee's impression that low-visibility, routine HHAs in particular do not systematically receive internal review at NEHC. The subcommittee recommends that a system for regular peer review of NEHC's HHAs by qualified internal and external reviewers be developed and documented. Such a system should include criteria for determining which reports will undergo internal or external review. For example, a report would undergo external review if it deals with a health hazard that might be encountered by thousands of Navy personnel and co-located civilians. HHAs of commonly used low-hazard chemical materials would bypass this external review process but would receive some internal review to validate the assessment of low hazard. HHAs associated with an intermediate hazard would be subject to systematic internal reviews.

The subcommittee believes that periodic external review of NEHC's HHA process would assure scientific rigor and objectivity of the program and provide an opportunity for staff to obtain additional perspective, overview and consultation, and access to evolving technologies from scientists working in the field outside the Navy. The subcommittee recommends that NEHC establish a peer review advisory board to assist it in carrying out its mission of providing credible and consistent HHAs. The board should be an independent body comprised of scientists with expertise in broad areas such as industrial hygiene, toxicology, epidemiology, and risk assessment. The subcommittee believes that the U.S. Army Center for Health Promotion and Preventive Medicine's (CHPPM) Peer Review Board on Toxicology is a good model and recommends that it be tailored to the needs and available resources of NEHC. The CHPPM peer review board was convened in response to COT's recommendation that such a review board be assembled as part of COT's review of the Army Environmental Hygiene Agency's (now known as CHPPM) Toxicology Program (NRC 1991).

COMMUNICATION WITHIN OTHER NAVY PROGRAMS AND WITH OTHER ORGANIZATIONS

NEHC receives requests from client commands for HHAs and these are promptly completed and returned to the client command in a timely manner. However, it appears that there is little or no interaction between NEHC and the client during the process and there is no feedback from the client as to the adequacy of NEHC's HHAs and recommendations. In addition, there were no examples provided of anticipatory or "proactive" efforts by NEHC staff to identify and mitigate potential hazards, or to develop improved processes for HHA. The subcommittee recommends that NEHC develop a formal system or process to collaborate with other military organizations, such as CHPPM in developing HHA reports and distribute report findings to them. Specifically, the subcommittee recommends that the following procedures be followed:

- Findings and recommendations of NEHC's HHA reports should be made available to all NEHC clients of HHAs and to the staffs of the health, safety, and environmental programs. In addition, the basic HHA information should also be made available to all naval personnel and civilian and contract workers, and to the lay public.
- NEHC's HHA reports should be entered into some form of an information system that is likely to be carried on an Intranet (restricted access) or the Internet (wide access).
- NEHC should establish and implement systematic customer survey and feedback mechanisms as to the utility and timeliness of NEHC analyses in decision-making, and obtain suggestions for improvement to better serve customer needs.

The subcommittee also believes that there is a need for greater coordination and information transfer between NEHC and other DOD, Navy, or governmental agencies that also perform HHAs. This interaction could provide insight for addressing issues and solving problems that might be common between institutions. NEHC would benefit by interacting more with CHPPM at Aberdeen Proving Ground, Maryland; the U.S. Air Force's Institute for Environment Safety and Occupational Health Risk Analysis at Brooks Air Force Base, Texas; Triservice Toxicology Research Laboratories (particularly the Naval Health Research Center's Toxicology Detachment) at Wright Patterson Air Force Base in Ohio, and the Armed Forces Institute of Pathology in Washington, D.C.; EPA; and the U.S. Department of Energy.

MEDICAL SURVEILLANCE AND CENTRALIZATION OF MEDICAL DATA

The Navy's Bureau of Medicine and Surgery (BUMED) collects medical data on Navy personnel. A centralized medical-data management structure is being developed for the entire DOD, which will incorporate data relevant to occupational and environmental health hazard assessment.

This centralized approach will also facilitate communication between various commands within the Navy and permit data entry from numerous remote sites (e.g., shipboard) with immediate feedback on quality control (e.g., valid response codes). It will provide NEHC with greater increased ability to access medical data to conduct surveillance to detect possible adverse health outcomes related to potential chemical exposures.

For the interim, the subcommittee recommends that data from the existing medical surveillance program be analyzed by NEHC to evaluate the effectiveness of HHAs on a regular basis. This is a key element of a properly functioning occupational health program, and is essential to verify the effectiveness of the HHA program and its recommendations. For example, data on the occurrence of adverse health effects in workers should be used to indicate need for additional toxicological testing.

LIFE-CYCLE ASSESSMENT

Although the DOD and Navy documents include policy statements requiring the use of life-cycle assessment (LCA), it is not clear to the subcommittee that LCA is considered and applied at all levels of the acquisition process, including development or purchase, use, storage, and disposal of potentially hazardous materials.

DOD Directive 5000.1 (Defense Acquisition Management Policies and Procedures) outlines a DOD acquisition process (See Figure 2-1) that is similar to many of the "phases and gates'" processes used in civilian industrial product-development processes. The HHA process requires identification of "potential environmental consequences" at Phase 0.

However, it is not clear that either occupational safety and health program costs or end-of-life considerations for various procurement alternatives are taken into account during acquisition as required by DOD and Navy policy. Both disposal and recycling of hazardous materials and hazardous waste, which are required elements of a complete LCA, are missing. The subcommittee recommends that questions related to these issues be considered in developing HHAs.

OVERALL SUMMARY

NEHC's HHA program performs a function that is basic to the occupational medicine and industrial hygiene programs of the entire Navy and Marine Corps. It is primarily designed to protect the health of naval and civilian personnel. In a reduced-size Navy, the preventive functions of NEHC can be an important factor in reducing costs associated with naval health care and readiness. At present, NEHC's HHAs provide advice only. The subcommittee recommends that the Navy elevate the importance of the HHA program by delegating authority to NEHC for "signing off" on decisions to use or not use products and by increasing support for the HHA program throughout the Navy command structure.

The subcommittee concludes that the development of formal, written SOPs; the addition of senior scientists with expertise in toxicology, epidemiology, risk assessment, and industrial hygiene; increased training of the current staff; better QA/QC procedures, including the formation of a peer review board; improvements in data acquisition and management; better coordination between NEHC and other organizations and stakeholders; the development of a centralized medical-data management system; and consideration of life-cycle issues would lead to a more effective HHA process that would stand up to critical and objective scrutiny.

The subcommittee recognizes that many of its recommendations will require significant resources to implement; however, the subcommittee believes that such investments to improve NEHC's HHA program would be cost effective and would result in significant and long-term benefits to naval readiness and mission fulfillment.

References

CHPPM (U.S. Army Center for Health Promotion and Preventive Medicine). 1999. Quality System, CHPPM Regulation No. 702-1. Dept. of the Army, U.S. Army Center for Health Promotion and Preventive Medicine, Aberdeen Proving Ground, MD (19 February 1999).

Clayton, G. and F. Clayton, eds. 1993. Patty's Industrial Hygiene and Toxicology, Vols 1-2., 4th Ed., New York, NY: John Wiley and Sons.

Dorland's Illustrated Medical Dictionary, 28 Ed. 1994. Philadelphia: W.B. Saunders.

EPA (U.S. Environmental Protection Agency). 1997. The Exposure Factors Handbook. Vol 1. General Factors, Report No. EPA/600/P-95/002Fa , Vol 2. Food Integration Factors Report No. EPA/600/P-95/002Fb, Vol 3. Activity Factors. Report No. EPA/600/P- 95/002Fc. Office of Research and Development, EPA, Washington, DC.

Ganak, G.M. 1998. Reengineering the MSDS process within DOD [abstract]. Presentation at the Fourth IBC Symposium on MSDS, Annapolis, MD.

IARC (International Agency for Research on Cancer). 1987. Overall Evaluations of Carcinogenicity: An Updating of IARC Monographs, Vol. 1-42, Suppl. No. 7. Lyon, France: World Health Organization, International Agency for Research on Cancer.

Keith, L.H. and D.B. Walters, eds. 1986. Compendium of Safety Data Sheets for Research and Industrial Chemicals, Part I. Deerfield Beach, FL: VCH.

Klaassen, C.D, M. O. Amdur, J D. Klaassen. 1996. Casarett and Doull's toxicology: The Basic Science of Poisons, 5th Ed. New York: McGraw-Hill, Health Professions Division.

References 59

Lewis, R.J. 1993. Hawley's Condensed Chemical Dictionary, 12th Ed. New York: Van Nostrand Reinhold.

Lewis, R., ed. 1996. Sax's Dangerous Properties of Industrial Materials, 9th Ed. New York: Van Nostrand Reinhold.

Lide, D.R. 1999. CRC Handbook of Chemistry and Physics. Boca Raton, FL: CRC.

NRC (National Research Council). 1984a. Emergency and Continuous Exposure Limits for Selected Airborne Contaminants, Vol. 1. Washington, DC: National Academy Press.

NRC (National Research Council). 1984b. Emergency and Continuous Exposure Limits for Selected Airborne Contaminants, Vol .2. Washington, DC: National Academy Press.

NRC (National Research Council). 1984c. Emergency and Continuous Exposure Limits for Selected Airborne Contaminants, Vol.3. Bromotrifluoromethane. Washington, DC: National Academy Press.

NRC (National Research Council). 1985a. Emergency and Continuous Exposure Guidance Levels for Selected Airborne Contaminants, Vol. 4. Washington, DC: National Academy Press.

NRC (National Research Council). 1985b. Emergency and Continuous Exposure Guidance Levels for Selected Airborne Contaminants, Vol. 5. Washington, DC: National Academy Press.

NRC (National Research Council). 1986a. Criteria and Methods for Preparing Emergency Exposure Guidance Level (EEGL), Short-Term Public Emergency Guidance Level (SPEGL) and Continuous Exposure Guidance Level (CEGL) Documents. Washington, DC: National Academy Press.

NRC (National Research Council). 1986b. Emergency and Continuous Exposure Guidance Levels for Selected Airborne Contaminants, Vol. 6. Benzene and Ethylene Oxide. Washington, DC: National Academy Press.

NRC (National Research Council). 1987. Emergency and Continuous Exposure Guidance Levels for Selected Airborne Contaminants, Vol. 7. Ammonia, Hydrogen Chloride, Lithium Bromide and Toluene. Washington, DC: National Academy Press.

NRC (National Research Council). 1988. Emergency and Continuous Exposure Guidance Levels for Selected Airborne Contaminants, Vol.8. Lithium Chromate and Trichloroethylene. Washington, DC: National Academy Press.

NRC (National Research Council). 1991. Review of the U.S. Army Environmental Hygiene Agency Toxicology Division. Washington, DC: National Academy Press.

NRC (National Research Council). 1992. Guidelines for Developing Spacecraft Maximum Allowable Concentrations for Space Station Contaminants. Washington, DC: National Academy Press.

NRC (National Research Council). 1993. Guidelines for Developing Community Emergency Exposure Level for Hazardous Substances. Washington, DC: National Academy Press.

NRC (National Research Council). 1994. Review of the U.S. Naval Medical Research Institute's Toxicology Program. Washington, DC: National Academy Press.

NRC (National Research Council). 1996a. Spacecraft Maximum Allowable Concentrations for Selected Airborne Contaminants. Vol.2. Washington, DC: National Academy Press.

NRC (National Research Council). 1996b. Spacecraft Maximum Allowable Concentrations for Selected Airborne Contaminants. Vol.3. Washington, DC: National Academy Press.

NRC (National Research Council). 2000a. Methods for Developing Spacecraft Water Exposure Guidelines. Washington, DC: National academy Press.

NRC (National Research Council). 2000b. Spacecraft Maximum Allowable Concentrations for Selected Airborne Contaminants. Vol. 4. Washington, DC: National Academy Press.

O'Donoghue, J.L. 1989. Screening for neurotoxicity using a neurologically based examination and neuropathology. J. Am. Coll. Toxicol. 8(1):97-115.

ORNL (Oak Ridge National Laboratory). 1989. Installation Restoration Program Toxicology Guide, Vols 1-4. Biomedical and Environmental Information Analysis Section, Health and Safety Research Division, ORNL, Oak Ridge, TN, for Harry G. Armstrong Aerospace Medical Research Laboratory, Aerospace Medical Division, Air Force Systems Command, Wright-Patterson Air Force Base, OH.

ORNL (Oak Ridge National Laboratory). 1990. Installation Restoration Program Toxicology Guide, Vol 5. Biomedical and Environmental Information Analysis Section, Health and Safety Research Division, ORNL, Oak Ridge, TN, for Harry G. Armstrong Aerospace Medical Research Laboratory, Aerospace Medical Division, Air Force Systems Command, Wright- Patterson Air Force Base, OH.

Proctor, N.H., J.P. Hughes and M.L. Fischman. 1996. Chemical Hazards of the Workplace, 4th Ed. New York: Van Nostrand Reinhold.

Quigley, D.R. 1994. Handbook of Emergency Chemical Management. Boca Raton, FL: CRC.

Risher, J.F., W.W. Jederberg and R.L. Carpenter. 1995. The assessment of health risk to occupationally exposed Navy personnel: A consideration of issues. Inhalation Toxicol. 7(6):983-1003.

Royal Society of Chemistry, 1994. Agrochemicals Handbook. Royal Society of Chemistry, Information Services, Unwin Brothers, Ltd, UK.

Wolff, R.K. 1997. Mechanistic Approaches to Providing Exposure Guidelines for Inhaled Pharmaceuticals and Other Chemicals. Abstract for Congress of the International Society for Aerosols in Medicine, Sendai, Japan, Sept. 22-26, 1997.

Appendix A

History of NEHC and Its Relationships With Other Navy Organizations

THIS SECTION describes the history of the Navy Environmental Health Center (NEHC) and its relationship with the Navy Bureau of Medicine and Surgery (BUMED), and other Navy organizations. NEHC originated in 1964 at the Navy Bureau of Weapons, which recognized the need for an occupational health program that would encompass all-fleet readiness and training ordnance field activities. The Bureau of Weapons directed the Naval Ammunition Depot (NAD), Crane, Indiana, to address this need by broadening the command's occupational health function. This expanded mission included providing assistance to all naval ammunition depots and naval stations in coordinating their occupational health programs and resulted in the completion of the first industrial hygiene survey.

From 1965 to 1967, the NAD program steadily expanded in response to new mission requirements for personnel education and survey procedures. In May 1967, the Bureau of Weapons formalized this function by establishing the Naval Ordnance Systems Command Environmental Health Center under the auspices of the NAD Crane Medical Department.

On July 1, 1970, the Center became a Headquarters Detachment of the Naval Ordnance Systems Command, which subsequently became the Naval Ordnance Environmental Health Center. In 1971, NEHC was brought under the control of BUMED and was renamed the Navy Industrial Environmental Health Center.

In July 1974, the Center became the Navy Environmental Health Center (NEHC), an Echelon 3 shore activity under the command and support of the Chief, Bureau of Medicine and Surgery. In the fall of 1978, the command was relocated to Norfolk, Virginia. This relocation was undertaken in response to an increase in requests for fleet support and further expansion of NEHC's occupational health mission. This expanded role included responsibility for the Navy Occupational Safety and Health Inspection Program, analytical laboratory services, radiation health, hazardous materials identification, asbestos hazard control, preventive medicine, epidemiology, and hearing conservation.

NEHC's mission was expanded again in 1981 to include coordination and provision of centralized support and occupational health, environmental health, and preventive medicine services to medical activities ashore and afloat. The Navy Disease Vector Ecology and Control Centers and Navy Environmental and Preventive Medicine Units were placed under NEHC as Echelon 4 commands at that time.

NEHC reports directly to BUMED, which in turn reports to the Chief of Naval Operations (CNO). BUMED is a large organization within the Navy with medical facilities on ships and ashore around the world. Because NEHC plays a central role in preparing health-hazard assessments, it is the primary source of information for occupational and environmental disease prevention programs managed by BUMED. The Chief, Bureau of Medicine and Surgery, is responsible for a number of activities that might require support from the NEHC. These activities include

- determining, validating, and establishing health-related criteria and standards that are not available through federal, state, or local agencies;
- providing assistance to activities, offices, and commands concerning the health aspects of pollution sources or pollution-control equipment, including development of medical monitoring programs;
- providing industrial hygiene and medical expertise for activities during spill events and other environmental emergencies via Navy hospitals and clinics, Navy Environmental Preventive Medicine Units, and Navy Disease Vector Ecology Control Centers;

- coordinating with the Agency for Toxic Substances and Disease Registry (ATSDR) for the timely completion of public health assessments for National Priorities List sites, toxicological profiles on any specific contaminants, health education, health consultations, and other activities provided in the DOD/ATSDR Annual Plan of Work.

In support of these activities, BUMED has issued directives that are in accord with current accepted standards of toxicological sciences and clinical medicine. Also, BUMED

- provides support in the areas of ATSDR public health assessments, review of toxicological profiles, environmental-risk communication workshops, review of site health and safety plans, and review of ecological risk assessments;
- coordinates with ATSDR concerning ATSDR's legally mandated health-related activities, including public health assessments, public health consultations, health surveys and investigations, toxicology databases, emergency response, Naval Facilities Engineering Command and health education;
- assists Naval Facilities Engineering Command (NAVFACENGCOM) and installations to prepare for public meetings and respond to community concerns regarding program health and safety.

BUMED also issues directives for shipboard medical personnel that provide guidance for determining, validating, and establishing health criteria and standards for chemical and physical exposures.

There is direct communication with, and support from, the Engineering Field Divisions/Engineering Field Activities (EFD/EFA) providing NEHC with pertinent risk- management options. Both EFD/EFA work with the Naval Facility Engineering Service Center (NFESC) and both report to the NAVFACENGCOM. BUMED and NAVFACENGCOM also coordinate their activities. Thus, there is an organizational framework within the Navy to detect and identify potential human and environmental health hazards, determine risk to health, and implement appropriate controls to protect the health of personnel. CNO has stated that, "The Navy chain of command must provide leadership and a personal commitment to ensure that all Navy personnel develop and exhibit an environmental protection ethic."

However, BUMED is the final arbiter regarding the prevention, diagnosis, and treatment of exposures to toxic hazards.

Appendix B

Department of Defense and Navy Directives and Regulations Relating To the Use of Hazardous Materials

TO COMPLETE its mission effectively, the Department of Defense(DOD), the Secretary of the Navy(SECNAV), Chief of Navy Operations, and the Bureau of Medicine and Surgery (BUMED) have published directives, instructions, and military standards. The purpose of this section is to review and summarize those documents, to assess whether the Navy management system for handling hazardous material clearly states its policy with regard to how such materials should be handled, who is responsible for implementing the policy decisions, and whether the Navy Environmental Health Center (NEHC) is adequately chartered to carry out the mission assigned to it.

Each of the relevant DOD or Department of the Navy (DON) documents are reviewed below.

DOD Directive 4210.15 of 27 Jul 1989: HAZMAT Pollution Prevention

This directive establishes policy, assigns responsibilities, and defines procedures for hazardous material pollution prevention. It is DOD policy that hazardous materials are selected, used, and managed over their life cycle, so that DOD incurs the lowest cost required to protect human health and the environment. The preferred method of doing this is to avoid or reduce the use of hazardous material. If use of hazardous material might not be reasonably avoided, users must apply management practices that avoid harm to health and the environment. Emphasis is placed on reduction of hazardous materials in processes and products, as distinguished from end-of-pipe management of hazardous waste. The managers of DOD components are required to (1) facilitate the use of a less-hazardous material when the use of a hazardous material or a process using a hazardous material has been authorized, and a less hazardous substitute is or could be available; (2) evaluate hazardous-material decisions using economic analysis techniques that match the magnitude of the decision being made, cost factors, and other intangible factors; and (3) begin economic analysis of hazardous-material decisions at the earliest possible stage of the life cycle and modify analyses whenever better information becomes available.

DOD Directive 5000.1 of 23 Feb 1991: Defense Acquisition

This directive requires a management process be used for acquiring quality products that emphasizes effective acquisition planning, improved communications with users, and risk management.

Threat projections, life-cycle costs, cost-performance schedule trade-offs, affordability constraints, and risk management are major considerations at each procurement milestone. A hierarchy of potential material alternatives must be considered prior to a decision to commit to a new acquisition program. Program plans must provide for a concurrent systems-engineering approach to achieve a careful balance among system design requirements, which include safety considerations. Project bid documents from contractors require them to identify risks and provide specific plans to assess and eliminate risks or reduce them to acceptable levels.

DOD Instruction 6050.5 of 29 Oct 1990: DOD Hazard Communication Program

This document establishes DOD policy, responsibilities, and procedures for a comprehensive hazard-communication program. Properly implemented, this program ensures that DOD personnel are aware of potential health hazards associated with their occupation; informed of safe work practices and proper use of engineering controls; trained in the selection, use, and availability of appropriate personal protective equipment to prevent chemically related injuries and illnesses; and comply with OSHA regulations.

DOD Instruction 6055.1 of 26 Oct 1984: DOD Occupational Safety and Health Program

This instruction establishes DOD policy that requires DOD units to establish and maintain comprehensive and aggressive occupational, safety, and health programs to protect all personnel from work-related deaths, injuries, or illnesses.

DOD Instruction 6055.5 of 10 Jan 1989: Industrial Hygiene and Occupational Health

This instruction establishes uniform procedures for recognizing and evaluating health risks associated with exposure to chemical, physical, and biological stressors in the workplace.

It is DOD policy to provide each employee with a healthful work environment that is free from recognized health hazards. DOD policy requires that health hazards must be identified, evaluated, and controlled.

Consistent, meaningful occupational health and environmental surveillance programs must be implemented to ensure that controls adequately protect the health of DOD personnel.

Military Standard 882 C of 19 Jan 1993: System Safety Program Requirements

This standard is intended to ensure that safety systems are included in technology development and designed into systems, subsystems, equipment, facilities, and their interface and operation.

Governmental agencies and contractors are required to use a safety-management approach during the system acquisition process and throughout the life cycle of each system, making sure mishap risk is understood and risk reduction is always considered in the management-review process.

It emphasizes a formal safety program that stresses early hazard identification and elimination or reduction of associated risk as the principle contribution of an effective system safety program.

SECNAV Instruction 5100.10 G of 15 Dec 1989: Department of the Navy (DON) Policy for Safety, Mishap Prevention, and Occupational Health Programs

This instruction aligns DON policy with DOD policy and stresses safety and occupational health as inherent responsibilities of the Navy command structure. Navy programs are to be established, funded, and maintained to protect all civilian and military personnel from work-related mishaps, injuries, and illnesses. Navy activities should emphasize an awareness of good safety and health practices among all personnel, both civilian and military. It requires that safety and health hazards be identified, evaluated, and controlled.

Consistent, meaningful occupational health surveillance programs are to be implemented by Navy medical departments to ensure that controls adequately protect the health of personnel. Personal protective equipment is to be provided. Safety and occupational health precautions are to be integrated into training and indoctrination programs and into technical and tactical publications.

This instruction also requires establishment of uniform procedures to evaluate safety and health risks associated with exposure to chemical, physical, and biological stressors in Navy workplaces. It also requires identification of safety concerns related to emerging technology, as well as the establishment and maintenance of a formal hazard-tracking system, to ensure that significant hazards identified during system safety program reviews are properly documented, tracked, and resolved.

SECNAV Instruction 5400.1SA of 26 May 1995:
DON Research, Development, and Acquisition and Associated Life-Cycle Management Responsibilities

This instruction establishes the duties and responsibilities of the Assistant Secretary of the Navy with regard to the research, development, and acquisition processes.

OPNAVINST 5100.24 A of 3 OCT 1986:
Navy System Safety Program

This instruction provides policy and requirements for the Navy System Safety Programs to improve operational readiness and reduce costs by using system safety design and analysis techniques.

Engineering and management controls are to be applied to ensure that prior to system production, construction, and deployment, primary emphasis is placed on the identification, evaluation, and elimination or control of hazards.

System safety risk requirements, criteria, and constraints, and needed program resources are to be addressed by the originators of each operational requirement and summarized in a Decision Coordinating/System Concept Paper. System safety hazard assessments must also be presented at design and program reviews.

Procedures must be developed for the safe and environmentally acceptable use, stowage, and disposal or demilitarization of any hazardous materials and equipment associated with the system.

Data must also be developed to identify and control hazardous materials and items, including selection of the least-hazardous alternative and provide safety and health requirements with the planned maintenance system cards, along with material safety data sheets.

OPNAVINST 4110.2 of 20 Jun 1989:
Hazardous Material Control and Management

This instruction establishes a uniform policy, guidance, and requirements for the life-cycle control and total-quality management (TQL) of hazardous material acquired and used by the Navy. It requires the Navy to identify hazardous materials needed to meet mission requirements and, if feasible, substitute less-hazardous material. It requires incorporation of the necessary investigations and research studies for safety, environmental protection, health-hazard identification, and risk assessments into system research and development programs. Assessments are to be geared to control and reduce hazardous-materials requirements, and minimize the costs associated with hazardous-waste generation and disposal.

The Navy is to control and reduce the amount of hazardous material used and hazardous waste generated by up-front hazardous-materials control in acquisition, procurement, supply, and utilization through the development of mechanisms to identify materials in the system that are hazardous, and to limit quantities of hazardous materials acquired and stored.

The instruction requires establishment of activity-authorized use lists, and controls over hazardous-materials quantities used to reduce the generation of hazardous waste. Plans for the review of specifications that direct use of hazardous materials are required to further minimize the use of hazardous materials.

The instruction also establishes mechanisms for substituting less hazardous material for hazardous materials if technically feasible. Decisions to use hazardous materials or substitution of less-hazardous materials are to be supported by an economic analysis appropriate to the magnitude of the decision being made. Such analysis is to include cost factors and intangibles such as savings from reduction in training and other related hazardous-material or hazardous-waste impacts.

BUMED, is identified as the office responsible for providing workplace-hazard evaluations and health risk assessments specific to hazardous-materials applications in the Navy. Also, the Navy is to develop, maintain, and distribute technical information on health risks and assessments for hazardous materials used in Navy workplaces and operations.

NEHC is to provide commanders and commanding officers with technical assistance in evaluating and monitoring the use of hazardous materials in the workplace, prescribing precautionary measures, and assisting shore activities in developing authorized hazardous-materials use lists.

OPNAVINST 5100.19 C: NAVOSH Program Manual for Forces Afloat

This instruction identifies the methods to be used to properly manage hazardous materials aboard surface ships and submarines.

NEHC efforts in recognizing potential health threats posed by hazardous materials are an integral part of this process. NEHC has recently begun a comprehensive program, in cooperation with the Naval Surface Warfare Center, Carterock Division, to become a major player in all decisions made about whether to allow a potentially hazardous material on ships, and inclusion in the Shipboard Hazardous Material List.

Requests by fleet and NAVSEA program managers for use of new materials must first receive NEHC review, along with line command consideration of the need any benefits of the new material.

An NEHC role in this process is to provide a comprehensive health-hazard evaluation report for each material, summarizing to what extent the material can be used safely aboard ships, and specific health-hazard control measures that will be necessary to ensure the safety of shipboard personnel.

NEHC also is instructed to develop a comprehensive Submarine Materials Review Program with the Naval Sea Systems Command (SYSCOM), so that all new materials contemplated for use in the construction or maintenance of nuclear-powered submarines can be reviewed by NEHC's Submarine Materials Review Board.

Reports issued to NAVSEA via BUMED recommend proposed-use categories, and contain additional health-hazard control guidance to ensure the material is used in a safe manner.

OPNAVINST 5100.23 D: Navy Occupational Safety and Health (NAVOSH) Program Manual for Shore Activities

This document addresses the Navy's Hazardous Material Control and Management Program. Responsibilities of BUMED include providing workplace-hazard evaluations and health risk assessments specific to hazardous-material applications, and developing, maintaining, and distributing to activities technical information on health risks and assessments for hazardous materials used in Navy workplaces and operations. BUMED is also responsible for providing commanders and commanding officers with technical assistance in evaluating and monitoring the use of hazardous material in the workplace, prescribing precautionary measures.

BUMED is to evaluate and confirm requirements for toxicological research for new systems or for Navy-unique hazardous material or Navy-manufactured hazardous materials. BUMED is to ensure development of needed data for the safe use and handling of the material in Navy systems, both ashore and afloat.

BUMEDINST 4110.1 of 30 Aug 1993: Hazardous Material Control and Management

This instruction establishes the policy, guidance, and requirements for the life-cycle control and TQL of hazardous material acquired and used by the Navy Medical Department.

It tasks NEHC to assist Navy systems commands, program managers, and medical department activities in support of the Hazardous Material Control and Management Program. This includes performing health-hazard risk assessments when requested by Navy systems commands, and involves the development, maintenance, and distribution of technical information on health risks and assessments for hazardous materials used in workplaces and operations, which is to be coordinated with Naval Medical Research Institute, Toxicology Detachment (NMRI/TD).

NEHC is to assist SYSCOM and program managers in reviewing hazardous-materials controlling documents (such as maintenance plans, maintenance requirement cards, and technical documents) that require the use of hazardous materials in the support, maintenance, or operation of systems and equipment.

NEHC is also to provide SYSCOM program managers with reviews of performance specifications and guidance on permissible exposure limits for the engineering control of hazardous materials in the workplace and to coordinate with the NEHC/TD to identify required toxicity studies and development of needed data.

SYSCOM program managers are directed to select nonhazardous materials or less-hazardous substitutes.

BUMEDINST 6270.8 of 6 Jun 1990: Procedures for Obtaining Health Hazard Assessments Pertaining to Operational Use of Hazardous Materials

This instruction is intended to minimize the health hazards posed by materials or by systems under development. It establishes formal procedures for obtaining toxicological information on materials in the research and development process being evaluated for introduction into the naval service or for new applications for existing materials.

It assigns responsibilities to NEHC for performing health-hazard assessments, and for publishing appropriate guidance for controlling potential occupational health hazards.

Programs involved in the research, development, and testing or evaluation are to forward requests for evaluation via their chain of command to NEHC. These requests are to be made early in the developmental phase of each program to allow sufficient time for the evaluation to be performed.

Information to be sent to NEHC includes a point of contact, details of the material or process, intended use, details regarding whether it replaces another material, plans for introduction and reporting deadlines for the evaluations, availability of funding to support any research required, and immediate notification to NEHC should adverse health effects attributable to exposure to the new hazardous material be documented or suspected.

NEHC is to provide interim responses to the requesting command to ensure obvious health hazards are identified early in the process, with recommendations for surveillance and control of hazards. Additional guidance is to be provided as results from further research become available.

Other commands (outside of research and development activities) must identify proposed new materials requiring an evaluation. New materials identified at operational unit levels are reviewed by the safety office of the operational command or by the regional Navy occupational medical department. If the information needed for assessment purposes is beyond that available to the local medical department representatives, the local industrial hygienist is directed to refer the request to the NEHC, where it will be evaluated.

NEHC is to be responsible for undertaking a formal review of new materials upon request and for providing an assessment of potential exposures, and the sufficiency of information to adequately characterize the risk provided by the new material. NEHC is to provide an interim assessment if there is insufficient toxicological information to characterize the risk presented by the new material.

This instruction includes recommendations for surveillance and controls, based on a reasonable and conservative interpretation and extrapolation of the available information. Required, but missing, data must also be identified.

NEHC is to coordinate with the Navy's TD in developing assessments for review of toxicological data, determining additional research required, estimating resource availability and project duration, and estimating additional resources required to advance such a project to meet the requesting command's deadlines.

NEHC is directed to disseminate information, as appropriate, on the hazards identified to ensure control of potential exposures and protect the health of personnel working with the new material.

BUMEDINST 5450.157 of 9 Feb 1996: Mission, Function, And Tasks of NEHC and Subordinate Commands

One of the functions outlined is to provide technical and professional consultative support and assistance to activities responsible for identifying, evaluating, monitoring, and correcting health hazards.